AF808017

Change From Within

The restructuring of the United Brotherhood of Carpenters and Joiners in Northern California - 1978 to 2008

By Michael S. Munoz

To:

Leslie

Pile Drivers, Divers Bridge Wharf and Carpenters Local 34

Phil Daniels, Thom Donnelly, Doug Downie, Pat Karinen, Bruce Lanham, Robert Lanham, James Lawson, Dan Murry, Red Poulsen, David Rhodes, Bob Smith, Steve Tilton, Bert Tycer, Larry Wagner and Jim Walker

John P. Sorenson

Archie Green

Roy Fouche

Jim McNamee

Editors:

Adam Machedo

Deborah Heinzmann

Cover:

Stuart Weinberg

Van Bourg, Wienberg, Roger & Rosenfeld

Table of Contents

Introduction: Atlantic City – 1991

The tension in the air was so thick you could cut it with a knife the day the votes were cast for the General Executive Board of officers at the 36th General Convention of the United Brotherhood of Carpenters and Joiners of America (UBC).

The Wardens appointed by General President Sigurd Lucassen were circling the voting booths like flies over a ripe pile of fresh manure and listening to the number of clicks coming from within. The voting machines had been set so that only one lever needed to be flipped to vote for the entire Lucassen Slate of candidates. Any other voting required each lever to be flipped. The wardens, by their proximity and attention, listening carefully, made every delegate fearful of the retribution that would come if more than one click was heard.

I had been asked by Whitey Rodgers to act as the campaign manager for his run against Sigurd Lucassen and their slate of officers. Upon reflection some eighteen years later, I continue to wonder what would have happened had Whitey Rodgers won. Would the UBC have changed drastically? I doubt it. Did the UBC change as a result of Lucassen's winning? You bet it did. Not because of Lucassen's great leadership; but because of the alliances he made to get re-elected and the fact that Doug McCarron, Executive Secretary of the Southern California District Council of Carpenters, was savvy enough to leverage himself into the spot of General President of the most powerful building trades union in the United States. As I stood with Doug McCarron, waiting for the ballots to be counted and the results announced, I noticed the concerned look on his face. This was a pivotal moment for the UBC and for us both personally. And as it turned out, it was historic for McCarron.

I had the backing of Jim Green, the Bay Counties District Council of Carpenters Executive Secretary. He was also one of the most powerful and respected Secretary/Treasurers in the UBC. Without his protection and support, I would not have survived politically. Reflecting back, I ask myself, have my experiences as a union organizer made a difference? I hope so. Could my experiences and observations of the difficult changes made by the United Brotherhood of Carpenters lead others in the labor movement to make the necessary changes in their unions and survive the next thirty years? And could these others rebuild the flaccid labor movement into a new, vibrant, and invigorating organization?

In 1994, after fourteen years of working in the field as a union organizer, I became Director of Organizing of the Bay Counties District Council of Carpenters (BCDC), a Union that only covered five of the Northern forty-six counties of California, while representing 60% of the membership and 70% of the hours reported in Northern California. By 2001, I was Director of Organizing for the restructured Northern California Carpenters Regional Council (NCCRC). Upon my retirement in 2008, I had seen Northern California go from 26,000 to 40,000 union members.

With this book, I hope to come to terms with the union's historical past and how I came to believe in the empowerment of working people and the belief that the labor movement will only survive with their involvement and participation. In addition, I hope to describe how a union's structure either facilitates it growth or precipitates its own demise.

A labor union as an organization is structured and managed. Ours, the NCCRC, became successful in an era of mass decline in union membership nationwide. The broad spectrum that this effort encompassed ranged from the local union operation to the structure of the highest international organizations within the labor movement. These changes in structure, operation, financing and organizing, working hand in hand, are key in the building of a new and modern union movement. As much I would like to say organizing is the key to success, having leadership with the will to change and the tenacity to see changes through is more important.

The labor movement is in a war of survival, and in a war, democracy suffers. I don't for a minute believe that electing business agents at a local union is the measurement for success or the model of what democracy in action should look like. Most union members are concerned with their conditions, contracts, and benefits. They rarely come to union meetings or participate in the day-to-day operation of the organization. Often this is no different than their participation in other aspects of civil government, i.e. local, state, and federal. Many of these non-participating union members have brought a negative reputation of unions to the workplace. The same arguments are used against nonunion workers when they try to organize. They may say that the union only wants their dues money, their representatives are all corrupt, they could make more money negotiating for themselves. A myriad of arguments are used to drive a wedge between union members and union leadership for the sole purpose of keeping workers' wages down, some members unintentionally aid the union buster with their negative gossip.

However, as part of the effort to rebuild and grow a modern, more efficient labor movement, union members have the responsibility to elect the person at the top, Executive Secretary or General President, It is the institutional memory of any organization that allows it to move forward without having to retool every few years. Unions are no different. Like apprentices, business agents require years of experience and training. Voting them out of office and replacing them with new and inexperienced representatives plays right into the bosses' hands. For their part, those who work for the union need to be accountable to a higher authority. Business agents must be held accountable for their actions, cronyism, trading jobs for votes, discrimination, or sexism. These are all grounds for termination, and should be. Accountability means jobs are dispatched fairly without discrimination. It means that the work will be policed and grievances will be dealt with before they become major problems. Accountability means organizers will be taught and expected to succeed. That success will follow the individual as he or she moves up into the organization's leadership positions, making the entire organization stronger, powerful, respected, and feared by the employers. Each one of us is only as good as his performance was yesterday. Every construction worker knows this and every organizer who has worked for me knows this.

This takes me to the AFL-CIO and its leadership to the emerging "Change to Win" labor organization, which is struggling for life. I have sat through conferences where the General Presidents of international unions have fallen asleep on the dais. You may criticize the Carpenters, but Doug McCarron is awake and ready to turn the work around or claim it from unions who are waiting for the government to change the rules and make their jobs easier. I believe that the enemies of labor, no matter how hard they try cannot put the "Labor Movement" to rest because the membership and many nonunion workers still believe in unions. That may sound odd during these times of decline but I continue to see workers standing up, putting their jobs on the line against unscrupulous bosses and corporations. I personally learned this in the fight against Rail Roadway/Hatch and the BE&K strikes. These nonunion jobs failed because the local unionized workforce that these employers were depending on to cross over and work nonunion, refused to work on nonunion jobs.

This isn't because of leaders like Richard Trumpka of the AFL-CIO or even Doug McCarron. It's because working people know right from wrong. They develop these ideas amongst themselves and act in their own best interests, in spite of the continual barrage by the media to highlight every fault in a particular union, or failure of an effort by a group of workers to organize. Unions are driven from the bottom. They exist because the members believe in unions, even if most don't even know of Richard Trumpka or any leadership above their most local level.

In 1996, I was called by Bob Lawson, who worked for the AFL-CIO, teaching the AFL-CIO Three Day Organizer Training. The Three Day Trainings were developed by Richard Bensinger and may well be the best thing to come out of the AFL-CIO in my lifetime. Trainings were designed to educate workers and identify leadership with the participating unions.

I was asked to teach a "Three Day" by Bob Lawson in 1995. We were to help the Association of Western Pulp and Paper workers (AWPPW) evaluate a group of potential organizers. I was teamed up with Cindy Chavez of the AFL-CIO, who would later be elected to the San Jose Board of Supervisors and run for mayor of San Jose. While the general president of the AWPPW was skeptical of the three day training, he was willing to try anything, and I believe we turned him around by the end of the training. In the class was John Rhodes who would become General President of the AWPPW.

In 1997 , having quit the Regional Council of Carpenters, Lawson wanted to offer me a job, and set up an interview with Mark Splane, Director of the Western United States for the AFL-CIO. I think I met with him four times before being hired by the AFL-CIO as State Director for California. I was asked to come to Washington D.C. for an interview with John Sweeney. I left the meeting feeling like I had just attended a Mary Kay convention, disappointed by the phony-feeling presentation that the labor movement was alive and vibrant. I quit three days later.

I wonder how much the decline in union membership is Lane Kirkland, John Sweeney or Richard Trumpka's fault? Union membership is in decline because the politicians who are responsible for protecting American Industry have sold out not only the working class, but the people of United States of America in general. The United States is the only industrialized nation with no "Industrial Policy" protecting American manufacturing. Also, many unions expect government to create laws that give all workers benefits for which unions have traditionally fought and bargained for. These Unions sit and wait rather than fight for the working class, while our conditions slowly sink lower and lower.

In the thirty-seven years since I first joined a union, I have often wondered why a wealthy person would run for office. I now realize politicians are not representing their constituents or even America; they are representing their personal wealth and the wealthy class. Whether Democrat or Republican, they represent Capital and the Global Economic Nations, those who have versus those who don't. Globalization has created a new political system that operates parallel to the democratically elected leadership we think is really running the world. But in fact our borders no longer exist. The United States is merely a state within the confines of the Western Hemisphere where money and goods pass freely, uninterrupted by fences, gates or physical boundaries. In the Global World Order democracies seem only to take up arms to protect Capital's investments.

Union decline is also directly measured by the industries who have abandoned this country and the corporate traitors who have sold America out. It is measured by the number of retirees who are no longer welcome as members in the labor movement, and by former union members who are not counted because they no longer pay dues to a union, yet support the principle of unionism. While membership is at an all time low, unions refuse to maintain a dialogue with these former members and millions of retirees who collect union pensions simply because they no longer pay dues. I contend that one is never out of a union; membership should be forever. If I believe I am a union member, dues payment should not be the sole criteria used to allow me membership or participation in a union or any organization, for that matter.

After quitting the AFL-CIO, I took my skills learned with the Carpenters and was hired as Area Director of Organizing for the International Association of Machinists and Aerospace Workers District Lodge 190. There I honed my skills in bottom-up National Labor Relations Board (NLRB) election campaigning. I joined my father's local 1546 and will be forever grateful to Don Crosatto Business Manager of IAM Lodge 1546 , who took the chance and brought me to meet with Directing Business Agent, Mike Day. .

If I have learned anything from my years of union experience, it is that the quickest way to lose an organizing drive is to try and do all the work yourself. Workers who want change have to also take responsibility, putting their jobs on the line and standing up to the management themselves. It is the organizer's job to get them to buy in and become union members before they ever pay a dime or officially join.

Any success I may have had has to be accredited to the leaders I worked for: John Watts, the Secretary/Treasurer of the Bay Counties District Council of Carpenters who spearheaded the fight for the dues check-off that started the modern organizing efforts in the carpenters Union in 1978; . Roy Fouche, the first Director of Organizing under John Watts; Jim Green, Secretary/Treasurer of the Bay Counties District Council of Carpenters, who restructured the Bay Counties Carpenters before any international intervention in a way that was palatable to the membership; Mike Day, of the International Association of Machinists and Aerospace Workers District Lodge 190, who gave me a chance and encouraged me, during a period where I could have walked away from the labor movement; Robert Alvarado, Executive Secretary/Treasurer of the Northern California Carpenters Regional Council, my friend and union organizer, who came up with me in the carpenters organization, and who I believe is the quintessential labor leader, the best of many who developed their skills under a system designed for him to succeed.

More than anything, I want this book to tell the story of a union that has grown and whose story can be used as a primer on how to rebuild the labor movement. And finally, why and how Pile Drivers Local 34 came to mean so much to me and my brothers and sisters in an era when other's unions are sometimes distrusted as much as government and politicians. Change is difficult; it can be defined as human nature, greed and goodness thrown together in a footrace on a path of unpredictability.

Chapter One: *My Family's Strange Origins*

While I am a natural-born citizen of the United States, my great-grandparents, along with 8,000 other poor Spaniards, immigrated between 1907 and 1913 to the U.S Territory of Hawaii on the HMS Willesden.

The plantation owners in Hawaii had gone through a series of labor disputes. The Asian workforce they had hired to harvest sugar were organizing themselves into unions, so the owners decided to bring in a workforce that could not speak Chinese or Tagalog, but could be directed by the Portuguese foremen. Company housing was segregated by race, partially so the plantation owners could prevent union attempts to organize along those lines.

I have always found it ironic that I was born an American because of a labor dispute, and that I would grow up to work as a union organizer. According to author Edwin Beechert in his book "Working in Hawaii", "These Spaniards moved to California almost immediately. Unlike the other races employed in the sugar cane fields, the Spaniards had brought their families and accumulated enough money quickly to move on as soon as their contracts were met."

To emigrate they willingly signed contracts binding them to work for five years. Advertisements distributed in Spain offered free land, payment in gold, jobs, and a hospitable climate where "The occasional Hurricane would whip you like an old man with palsy". Hawaii must have seemed attractive to these immigrants. My father, Manuel Munoz, was born at the plantation of Ewa Mill on the Island of Oahu in August, 1921.

By 1930, most Spaniards had moved to California, found work and assimilated into American society so smoothly that very little has ever been written about these experiences. My family tree has gone from surnames like Munoz, Campos, Yglesia, and Sanchez to include non-Hispanic names like Adams,

Advertisement – Spain 1911

Brooks, Raffanti, and Promes. We are now a family of every race and ethnicity, reflecting all of American society. But despite all the assimilation, my family takes great pride in our Spanish blood. My father used to say, "There was a time when the sun never set on the Spanish flag". Of course this was largely accomplished by murdering everyone who got in their way.

My mother's grandfather, Juan Alarcon Sanchez, was drafted into the Spanish army in the 1890s and served twelve years as a draftee in Spanish campaigns in Melilla, located in North Africa, and in Cuba during the Spanish-American War. In stories told to me while collecting and building our family tree, and through translating his army dispatch papers, I learned that he had been in combat during most of that time. I also learned that he and many other Spanish families chose to immigrate to Hawaii because their 18-year-old sons were becoming eligible to be drafted into the Spanish army. The Spanish army was harsh and disciplined towards soldiers and recruits, separated by a caste system of officers who came from the privileged Spanish society and nobility.

An anecdote highlighting this harshness was told about General Franco around the time he took command in North Africa as a Major. There was a mutiny by the soldiers against the inedible food they were forced to eat. The soldiers rioted, breaking beds and smashing the barracks. The men were called to assemble before the General with their mess kits at hand filled with the slimy mess they were expected to eat. As they stood at attention, one solider put out his mess kit, and a small amount of food splashed on the General's clothing. The legionnaire stated, "Do you expect us to fight on filth like this, Major?" Franco brushed off his uniform and ordered the captain in command to bring a plate of the questionable food. After tasting it he ordered the Captain arrested and deemed the food to be unfit for human consumption. Franco ordered the cooks to throw the food out and re-cook the entire meal. When the meal was completed he ordered the men to attention, called forward the solider that had spilled the food on his shirt and had him shot.

Upon discharge, Juan Sanchez moved to San Roque, eleven miles from Gibraltar. He would buy pots, pans, cloth and other sundries and take them in his wagon and trade with the farmers for eggs, vegetables, cheese and meats. His wife Margarette would then sell them back in town. It was a very hard life.

My father's grandfather, Andres Corrales, was a campeseno, a peasant who lived in a house in Gaucin, Andalusia with dirt floors. His wife, whose family owned a cantina, disowned her for marrying my great-grandfather. He died on the island of Kauai, Hawaii, and was buried there.

The emigration from Spain was not aboard a passenger liner; the HMS Willesden's normal cargo was not people. My fathers mother Isabel told me that they were all sick and lived on macaroni. People who were too sick were disembarked in Cuba and Argentina. My Grandmother Rose Campos told me how she had been riding on a small train through the sugar fields in Hawaii, when the train crashed. She had fallen clear, but her girl friend had been cut in half and killed.

My father's mother Isabel Corrales married Manuel Munoz in Hawaii in 1920. She said he put a spell on her, but the spell was broken and she divorced him in the early 1920's. This was not something Spanish women did in those days. She married Cayetano Ciurana in 1927, a Spaniard who I remember as my grandfather and who my father thinks of as his father. We have pictures of my father and uncle Andres Corrales, my grandmother Isabel's brother, posing for a photograph in Westwood, Ca. My grandfather and uncle worked for a short time for the Feather River Lumber Co. Later Cayetano would become a member of the molder's union in Oakland. He shoveled 20 tons of sand a day in the foundry alongside his brother in law Edwardo Yglesia. Cayetano was taught by his brothers to read and write, while cutting wheat with a scythe. He kept a diary of important events and was rumored to practice witchcraft and be able to cast spells. He died in his driveway a few weeks after his retirement. I remember him for his happiness, kindness and patience.

My grandmother Isabel and Cayetano were typical in my mind of Andalusian people; everything they did seemed joyous and positive, they loved everyone and lived life to the fullest. They drank and smoked and sang flamenco music until late at night. For my father's family the sun was always shining. My love of family and people was inherited from them. In my heart and dreams Cayetano Ciurana is with me every day.

My mother's family were from Vacaville California, and very different than my father's side. Grandmother Rose's family were typical Spaniards, social and gregarious, especially with family. My grandfather's family was more reserved and isolated themselves. Jose Campos, my grandfather, was a hard man and very old fashioned. He ruled the roost, and my grandmother Rose served him. When she was eighteen, my mother eloped with my father following World War II, because her father did not approve. My oldest memories of my grandfather Joe are that he wore a suit and a hat whenever he left the house, and in one pocket of his coat he carried a gun, in the other a pint of Four Roses Whiskey. He was a union boilermaker who worked in the Kaiser Shipyards. He died in his 50's from stomach cancer and as time passed we would go to St. Elizabeth's Catholic Church in Oakland with my grandmother and light candles and pray for his soul. She must have thought he needed it.

My grandmother Rose Campos was member of the Teamsters Union and retired after working 25 years at the Del Monte Cannery in Emeryville. She was very independent after my grandfather Joe died. She was also very kind, and raised my two cousins Mary and Annie Sanchez who lived with her along with my uncle John Sanchez for many years.

My father Manuel is a hero to me. He is a U.S. Army Veteran, a survivor of the Normandy Invasion. Like every other male member of my family alive during WWII he defended the United States. He was with the troops who landed on Omaha Beach on D-Day, at Saint Lô, one of the great battles following D-Day; and he marched in the liberation of Paris.

As a diesel truck mechanic, he could fix anything. I can't remember him driving a car that he hadn't bought from a junkyard and rebuilt. He was a member of Machinists Lodge 1546 in Oakland CA. One of my father's greatest gifts to me was the time he spent teaching me about politics, government and power, how the government isn't always truthful. Hours of discussion and argument with him at the kitchen table taught me my idealism, love for the underdog, distrust of politicians, and love for the union.

My mother Dolores, to our family, was the rock of Gibraltar. Sisters Linda, Wendy, my brother Robert and I grew up in the era where women did not work. My father took great pride that she stayed home and raised the four of us. But, every summer my mother, my grandmothers, aunts and cousins, would head to the Del Monte Cannery in Emeryville California, to work the fruit season as members of the Cannery Workers Union.

It was always clear who we were, that we belonged to a community of Spanish people. At every turn in my upbringing I was loved, by my parents, my grandparents, uncles, aunts and the friends they surrounded themselves with. Men kissed each other in clear displays of masculine affection. They sang and danced anytime they got together. We were never taught to hate or to believe we were better than someone else, for that you got a spanking. And they all belonged to a union.

For all of the family unity with regard to unionism, as I grew older I noticed there was a difference between the men and women in our family with regard to their relationship with the Catholic Church. The women attended to our religious needs, making sure we were baptized, that we attended catechism, had our first Holy Communion, and were confirmed by the church; from the men, zip, zero, nada. My Uncle Blas told one story that sticks in my mind. During the Depression my grandfather Joe Campos had gone to the church to ask for help. The priest gave him a dollar and told him never to come back. So, he didn't. Another story I discovered when collecting my family genealogy in the 1970s was told by my cousin Frank Bera of Vacaville. During the Spanish Civil War, Frank's father Francisco was active in raising money for the anti-Franco forces. He also sent money to one of his brothers who attended a Spanish seminary and became a priest. Another of Francisco's brothers, Peter, was a soldier on the anti-Franco side. Peter was captured, and Francisco wrote his brother the priest and asked if he could help secure Peter's release in exchange for Peter immigrating to the United States and never returning to Spain. The Catholic Church supported Franco. Unfortunately, Peter Bera was never heard from again; no one knows what happened to his wife and children.

All of these stories, of my great-grandfather's war experience, his desire to make sure his sons were not drafted, the perceived lack of support of the Catholic Church by my grandfather during the Depression, my cousin's losing effort to save his brother, perhaps exacerbated by a politicized Catholic Church, have helped strengthen the ideological foundation my father built at the kitchen table. Always questioning and being wary of concentrated, organized power in any form, whether it be political, governmental, or religious.

In my heart of hearts I believe that government is a necessary thing, and democracy, especially at the local level, is essential to keep communities safe, organized, and flourishing. However, at the federal level, government appears to be a tool of the rich, the churches and conservative elements in the world, used to reserve power and wealth for a small minority, keeping the poor and the working class struggling and therefore in order, toiling to produce more for them. Well, if the powerful eat steak, I want a sandwich.

Chapter Two: *San Leandro, CA*

The Bonaire District in San Leandro, California is right off highway 880, west on Marina Blvd, make the first left on Merced and follow it into the old neighborhood. This is a working class suburban area; you'll find them all over America. Our fathers worked, our mothers stayed home. If you did something wrong, somebody else's mother or father would straighten you out, or go have words with your parents.

We played sports from sun up to sun down, whatever sport was in season, and others played in a band. Music and sports: what a life. Our fathers were coaches and, for the most part, World War II Veterans. There was not much whining or sniveling allowed. We were taught to be tough and honest. Our fathers were all working class, tough minded men who taught us to appreciate what they had fought for.

During any summer day at eight in the morning at James Madison Elementary School two baseball teams would be picked from the assembled crowd of kids. Usually there were more than eighteen kids. So we even had a built in crowd to watch and replace those whose mothers would take them away at any given point.

Picking sides in a sandlot baseball game is the essence of democracy. The kids assembled, through a loose convention of delegates, decide who will be captains. Once done, the captains flip a coin or a baseball bat is thrown in the air handle first, each captain takes turns gripping the bat until ones hand is closest to the handle. He wins and chooses first.

The captains, generally the most liked and best baseball players present at the time, would take turns selecting a player at a time until a full nine-man team was selected. Those not chosen suffered the humiliation of being deemed not good enough by their peers. None of that unrealistic concern over creating low self esteem, my-child-is-fragile bull crap. You're good enough or you're not. Move on, get over it, play ball. Like a construction site, or any shop.

In this small area, probably ten square blocks, after high school, a few of us chose another life path, not college but the trades. Three men I grew up with would also find themselves working as pile drivers, carpenters and machinists. All rank and file journeymen, they would have a profound effect on how I made decisions that affected their lives when I went to work for the Union. Not in a way where I was looking after them, but in how they directed my decision-making based on their daily experiences in the field as union members.

I have come to believe that people who play sports when they are young become excellent construction workers, their agility, coordination and experience practicing and developing skills through repetitious practice make them great mechanics.

Mike and Pat Carvalho were always there. Mike was my age, and always either a captain or picked first. In 1970, Mike would be awarded a Bronze Star and Combat Infantry Badge fighting in the A Shau Valley during Operation Firebase Ripcord in Vietnam. He became a foreman parts man in the trucking industry and active union member in the Machinists Union. And someone I could trust to give me an honest opinion.

Pat was three years younger, smaller and always picked in the first round. Pat was one of the best athletes to come out of the neighborhood. Baseball, football, basketball, he played every sport well, but he excelled at football.

In 1975 his football career ended at Cal State University and in 1978 found his way into the Carpenters Union Apprenticeship Program at local 745 in Hawaii. One day I received a call from Pat. He had moved home and was working for a company called Tilden Coil. They were remodeling the Masonic Home in Fremont, California. His employer was signed to the union agreement in Southern California but not in Northern California. Pat asked me what we needed to do to organize the company. I told him he would have to take some risks and do what I asked of him. I showed up one morning by myself with a trunk load of picket signs. Pat had prepared everyone to walk off the job and within days the company was signed to the Union Master Agreement.

Don Dolly lived in the same neighborhood. He was ten years younger than I, and like Pat Carvalho, Don was an athlete. He played football and went off to the University of Nebraska on a football scholarship. He was an avid weight lifter, and while at the gym one day met Terry Price, a pile driver and close friend of mine, who lived next door to my parents. Over the following weeks Don and Terry became friends.

As he had done for me, Terry Price inoculated Don on unionism and the greatness of the Pile Drivers Union. Don's father was also a strong trade unionist and encouraged him to join the union's apprenticeship program. Don completed the four year apprentice program, working at any and every job he could. Following in Terry's footsteps he ran for office and was elected Warden on the Executive Board of Local 34 while I was President. As a journeyman pile driver he was chosen to be the steward on the job by the men. He became in the truest sense of the word a journeyman pile driver.

He became a foreman for Foundation Constructors, one of California's largest pile driving contractors, taking pile driving crews all over the Western United States. Don was offered stock in the company and he borrowed against everything he owned to buy a stake, and during the last few years has served as President of the Pile Driving Contractors Association.

Don is a great example of a company owner; he is not unique in Local 34. Like many company owners Don served, as an apprentice, he worked to become a journeyman, then as a journeyman represented the men as a steward and union officer. He then became an owner. Don understands what is being asked of the men, because he has done what he is asking of the men. He recognizes the men as stakeholders in the success of the company. A good crew can work together for years, making money for the company and staying employed throughout the year. The more they make the more crews are hired, everyone making a profit.

What makes Local 34 unique is that many in management have been rank and file pile drivers. They can see dangerous situations before they exist. Their expectations are based on real experience, not theory. They accept the union agreement as part of the daily business because it works. They have been hurt by the work, injured long before they became superintendent and owners. Their first-hand experience with injury and death is real.

They know that the union and the men are the same. A union member will enforce the union agreement, if he can't, he relies on the union steward to enforce the rules or call the union agent to enforce the agreement. Solutions to problems can be reached, the job goes on, and the men go on to the next job and most importantly they come home from work at night, safe.

Don Dolly, Mike and Pat Carvalho, three athletes with tools, found their niche becoming pile drivers, carpenters and machinists. Coming up from the playgrounds in San Leandro, like many workers, they all found a profitable use for their athletic skills. They came from working class families and working class neighborhoods.

Because of their working class ideology, I have always relied on and respected their opinions of the union's contract, their perceptions of the men they worked with, their ideas on what would improve the union and their overall view of the industry. They take great pride in their trade and the physical challenge of the profession. To them the union is not a building or a contract that controls their lives. Union membership is more of a team effort, the kind learned in athletic competition and reinforced through practice and the sprit born of pulling together for the collective good.

Chapter Three: *Pile Drivers Local 34*

In 1972 I met Terry Price; he lived next door to my parents with our neighbor Agnes Battaglia. Terry was a Marine, a Vietnam Veteran, and the most gregarious person I had ever met. We spent hours talking about the Vietnam War, the unjustness of the Nixon Administration, and about unions. He could swear for ten minutes, insult a person from head to toe and never repeat the same vulgarity twice.

I knew only what my family had taught me about unions. I was working for the Cyclone Fence Company and making a lot less than the union iron workers I met working for the same company. To my advantage my father had drilled it into my head that, someday, when I grew up and realized I was part of the working class I would drop the nonsense and quit college. I had been attending Chabot College in Hayward, California, but I was losing interest and only there to play soccer. My father was hoping I would find a real union job, learn a trade and move out of the house, and become a productive citizen.

During this time Terry joined Pile Drivers Local 34. He would expound on the greatness of the union and the toughness of the work. Only a few exceptional tough and intelligent human beings, he said were qualified to become a pilebutt. A term pile drivers use to refer to each other. In 1974, President Nixon froze wages for all Americans. The pile drivers and carpenters went on a wildcat strike to protest this decision. They had fifty cents per-hour coming and they wanted it. I went with Terry to union meetings at the Building Trades Temple on Valdez Street in Oakland and watched with amazement as construction workers went out everyday and shut down jobs. These weren't college students or longhaired hippies. They were rather large conservative men who were pissed off. I was sold; I was going to become a pilebutt if it killed me. Little did I know how close it would come to doing just that.

I had been going to Pile Drivers Local 34 every day trying to find work. My name was on the "C" list, which meant I would get a job off the list after Jesus came back from the dead. There was an "A" list you had to be qualified to be on, a "B" List for those of lesser status and finally the "C" list for those of no status, knowledge, or family to help them get work. Since then the "C" list has been eliminated and, to the benefit of all want-to-be pilebutts, an apprenticeship program was created so people like me would be trained and taught how to do the job well and safe. I did go before the Local 34 Examining Committee and was approved by S.W. Price, a very formidable looking man and a well-respected pile-butt. I put my name on the "C" list, Joe Lumsden the Financial Secretary looked at me through the dispatch window everyday and rolled his eyes in discouragement. But one day there appeared on the dispatch board a job for a "cutoff man". I told Joe I wanted to cut off pile and could operate a jackhammer, and he looked at me like I was Jesus who had just come back from the dead.

One of my jobs working for Cyclone Fence was to tow, maintain and operate a jackhammer. One of the hardest jobs in pile driving is pile cutoff. After a piling is driven into the ground it has to be cut off at the grade point, where the building starts. Concrete piling had to be cut off using a jackhammer, which is hard work in itself. But the jackhammer has to be held at a ninety degree angle to the piling, which means you have to hang it from a piece of rebar lodged in to one of the holes in the top of the pile, or you have to hold it in your arms. I was a jack hammer guru, so I thought.

There just happened to be a cutoff job at the Southern Pacific Building in San Francisco and I was dispatched out to work that day. When I arrived, I went down into the hole where I met the largest man I had ever seen; Emerson Franks, or "Franks" as everyone called him. To this day I remember him saying to me, "Sonny, you just keep up with me and you'll do fine". My job was to ring the pile, which meant using a chipping gun to chip a line around the grade mark so that when the jackhammer began breaking the pile, it would not chip or break lower than the line I had chipped.

I thought, it would be a cinch, Franks was at least 60 years old and he could do it. I was going to be making $9.45 per hour as a journeyman pile driver and not have to work as hard as I had been for Cyclone Fence. Unfortunately, I was underestimating Emerson Franks. He worked me into the ground, and by the end of the week my hands were bleeding even though there were no cuts on them. Still, from the first day I walked out on that job, I knew I would make it no matter what.

We were working six days a week, ten hours a day. I received my first paycheck on the third Thursday of the month, union meeting night. Red Paulson the steward, told me to be at the union meeting and that I was expected to make good my initiation and dues. They weren't going to let me back on the job unless I was good with the union. My whole paycheck went towards Initiation Fees. I went from being rich to poor within a few hours.

That job must have had every character in the union, Red Paulson, Bill Keller, Sam Keller, Bill Hamilton, Emerson Franks, Shane Keller, Charley Gibson, Arden Searls and many others. Every night after work we went to a bar on Stuart Street, it was always full of pilebutts, jitney drivers and stevedores. I quickly found out that all the pilebutts drank for free. It seems that the scrap metal on the job was going straight to the scrap metal yard and sold. I later found out that some of the pipe pile was going straight to the scrap yard as well. It was a dream comes true; I was earning more money than I had ever made and the booze was free.

One day, Bill Keller, one of the foremen on the job, and about as mean as any man I had ever met, went to the bar at lunch and got into a friendly fist fight with one of the stevedores working a block away on San Francisco's Embarcadero. The stevedore went to his union hall and found a bunch of friends and came back to the job site to "kick some pilebutt ass".

Shane Keller, who was about my age, locked everyone in the site and the fight was on. I looked up from the hole where I was working and saw it going down. During those days I had experience in this kind of fun, so I began to crawl out of the hole to help my new found brothers. Emerson grabbed me by the seat of my pants and said, "Son, you stay down here with me. Those are mens [sic] fighting." Those poor stevedores must have thought there were five or six guys to beat up. I'll bet thirty pilebutts came out of the hole like a swarm of angry hornets that day: Pilebutts 30 - Stevedores 0.

I later found out that there might have been another reason for Emerson's restraint: he had been convicted of murder and served time in San Quentin when he was younger for killing a man in a domestic dispute. I was learning that in this environment, a man stood up for himself and for his friends.

I went to every union meeting after that and learned about workplace democracy, defending a position, standing for what is right, fighting for a just cause and always being loyal to Local 34.

Like all construction jobs our work on the Southern Pacific Building came to an end. My next job was for Peter Kiewit & Sons at the Standard Oil Refinery, now known as the Chevron Refinery in Richmond. Not being related to anyone, I was given sound advice by Emerson Franks when we said goodbye; if everyone is standing around bullshitting make sure you are working. And that's what I did. I kept moving and the harder I worked the more pilebutts helped me learn.

It was on this job that I met my lifelong friend David Rhodes, one of the bravest men I have ever known. I had been directed to climb to the top of the leads to grease the shivs. The shivs hold the cables that carry the load of the pile driving hammer and other necessary cables used to drive piling. I was expected to climb one-hundred and ten feet straight up a ladder with a grease gun and apply grease to the shivs. This was the first time I had ever climbed so high, about seventy five feet up; my muscles began to burn and I stopped to rest. Next thing I knew David Rhodes was ten feet below me asking me if I was all right. I told him I was just tired and resumed climbing and finished what I had been instructed to do. The next time we met I would be in the hospital.

I was 24 years old and had made it as a pile driver six months, and was in the best condition I had ever been in my life. I drove pile all day, played softball at night and soccer all weekend. Then one day I was assigned to help tear down the crane so that it could be moved to another site on the Standard Oil property. While dropping the spotter, (a hydraulic machine that moves the leads in or out), the operator was bullshitting with the foreman, leaned on the lever and accidently closed the spotter without looking first to see if all was clear. I happened to be standing in between the spotter's bracing as I climbed down. I saw, I heard it close; I lifted my right leg out, as the machine closed on my left leg. I heard my leg splintering and then snap. I heard someone scream. Later I was told it was me.

The bones in my left leg had pierced my boot; the upper tibia and fibula went out the left side, the lower bones went out the right. When the operator opened the machine I fell down onto the spotter. Vern "Corky" Houk was immediately next to me, with my blood squirting into the air I asked if my foot was still there, "Naw," he said, "It's just a scratch. You'll be okay." He stayed with me and got me into the man lift.

I awoke from surgery to find my leg in a cast up to my thigh. A woman in a flowered dress was standing in my room. She asked me what kind of drugs I liked to take on the weekends. I found this strange and told her I didn't take drugs, I was a soccer player. The nurse came in and said that since the woman was not a relative, she would have to leave. But that incident scared me and I began to realize other circumstances were at play. I had been hurt because the foremen and operator were not paying attention to the work at hand. The company needed to shift the blame on me to limit their liability.

I also remember being visited by the Kiewit superintendent, who said, "Boy, you really screamed up there." It was as though I should have been embarrassed for screaming when my leg was being crushed in a hydraulic ram. I feared that his only concern was for the company and the placement of blame on me.

It was Pile Drivers Local 34 who came through for me, helping with my benefits and legal advice. It was also where my friendship with David Rhodes developed. David had been working with me on the job site and was the first person to visit me in the hospital. His concerns for my well being led to a lifelong friendship that would later evolve into a partnership in the fight against the Carpenters International Union.

I was now on disability. My leg was in a cast for one year and a brace for another year. I have to say that this was the most horrible thing to ever happen to me. But I swear that for every bad thing that has happened something just as good offset any accident I may have had. During my recovery, I went back to school. When I had previously quit junior college I was one English class short of graduation. My grade point average was less than 2.0. I quit college just before they kicked me out. Now I needed an "A" to receive my AA degree in General Education and guess what, I studied and got an A.

I had wanted to be a pilebutt, and I learned that working while others were standing around had served me well. Learning something the first time was also a good thing. Getting an A in college was not as hard as getting Ds. I worked at getting healthy and was determined to get back to work, no matter how hard it would be. I could do anything I applied myself to. However, playing the game I loved the most, soccer, was over for me. While recuperating I was asked to coach a youth soccer team. I would later coach a Under 14 boys youth team with friends in Livermore, the Clippers, that would go on to win the state championship. Coaching athletes would become teaching organizers and stewards later on in my life.

In 1976 Jack Melton, the business agent for Local 34, called me at home. He said the union was having a problem on a job in Oakland and asked me if I would walk picket. I said sure, and showed up for picket duty at the old Rhodes and Jamieson yard at the foot of 29th Avenue on the Oakland Estuary.

29th Avenue was also known as "Jingle Town", named for the Zoot Suiters of the 1940's who would jingle the change in their pocks as they strutted about town. The neighborhood had been predominantly made up of Portuguese Immigrants. The City of Oakland was trying to redevelop the area and had hired a non-union contractor to rebuild a dock on the Rhodes and Jameson's property located next the Park St. Bridge. My job was to stand at the driveway. I was instructed by Jack Melton to be there at five a.m. and only give out his phone number if anyone asked, other wise keep my mouth shut. By five thirty the Teamster began to arrive for work, not one crossed the picket line. By ten in the morning the work and been let to a union pile driving contractor, I was scared and elated at the same time, and sold on union power.

Chapter Four: *Hardknocks*

Once my leg began to heal and the cast had come off I was fitted with a knee high brace, which I wore for another nine months. I remember how difficult it was living with it, and I recall how differently I was treated by whoever noticed the brace. I worked hard at appearing normal.

I learned a lot about the worker's compensation system through this ordeal. When taken off worker's compensation by my doctor he told me to go back to work on light duty. There is no light duty in pile driving, and working while wearing the brace was torture.

As kids playing sports we used to tease each other when someone got hurt, asking the injured person "Does it hurt?" or "Tape it" meaning, suck it up; no pain could stop us. Since there is no "light duty" in pile driving, I thought to myself, "...fuck'em, I'll show'em." And I went looking for work.

One of the guys from the Southern Pacific job, Don Pruner, hired me to cut off pile. Pruner was a great pilebutt, and was almost always drunk. It wasn't until the fourth day that he noticed the brace on my leg. I had been carrying the jackhammer up and down a sixteen foot ladder when Pruner and the superintendent saw the brace. The superintendent was amazed that I could cut off pile with a brace on my leg, and marveled at how I could climb the ladder carrying the ninety-pound jackhammer on my shoulder. He also told Pruner to lay me off...so much for that display of strength, dexterity and pain tolerance.

I wasn't going to let that happen again. I had to build my strength back up, the brace had to go. I couldn't run so I went to the Franklin Street Boxing Gym in Oakland, and began working my legs to build enough strength so I could work without the brace. I believe that construction workers are like athletes, that good conditioning and the right mental attitude go a long way in making it a career.

Pier 39 – San Francisco – Photo by John McGuinn

Soon I was strong enough to work. In 1976, my friend, Terry Price, had found a company that was keeping him working pretty steady. They had a job coming up in Lovelock, Nevada and he helped me get a job working for Jenson and Reynolds for almost a year driving pile. I can't remember working harder, traveling farther from home or having so much fun. During the summer of 1977, I went to work for Santa Fe–Pomeroy on Pier 39 in San Francisco. Our job was to remove the old fender system (to protect the hulls of ships tied alongside the dock) to make room for the pile driving crew to enlarge the dock for a shopping center and boat harbor.

While we were doing our job, the Cleveland Wrecking Company was getting ready to demolish the warehouse still standing on the pier. They planned to finish the demolition on Saturday when everyone else was not working. They had been removing the roof, first by taking out ten joists and leaving two all the way down the length of the 1,000-foot-high, 25-foot-long building. They had also moved a crane inside so they could remove the debris after the wall was dropped. They had successfully performed this move on the west side of the warehouse a few weeks earlier.

That Friday, we had noticed that the wall vibrated and were concerned that the crane inside the building was rubbing against the wall as they prepared it for the demolition on Saturday. Our foreman, Gary Bakke, told my good friend Jim McNamee and I to keep awake and know where we were going to jump if something should go wrong. Knowing where to jump is something every pilebutt learns the first day on the job.

Around 3:00 p.m. the wind began to blow like it always does in the afternoon on the San Francisco Bay. It blew from the west right through the warehouse and against the inside of the eastern wall. At the same time the crane operator inside swung the crane and the counter weight brushed against the inside wall.

We were working on the outside of the pier shed when I heard a great tearing sound; I ran and dove off the dock. I remember seeing Jim McNamee's eyes as he dove with me into the bay. I briefly thought how embarrassed I would be if the wall was not falling. As I left the dock I first felt the wind created by the falling wall and was struck by rocks being blown off the pier. Then, while I was still in the air, the wall slapped me on the right side of my body. I was propelled eighteen feet down to the water, peeing in my pants out of fright. Hitting the water felt as though I had done a belly flop and the wind was knocked out of me. I was sucking sea water as I inhaled. During this time I was awake and could see the surface, but my life jacket wasn't bringing me there. I told myself not to panic and used the quick release, cutting my tools loose.

I hit the surface choking. Jim McNamee was right there to grab me. Gary Bakke had dived under the front of the crane and the wall had stopped within an inch of crushing him. He had a nail in his shoulder but was otherwise uninjured. I watched with amazement as pilebutts, some I knew and some I didn't launched themselves off the dock to help Jim and me. They had no idea what danger might have been in the water; someone needed help and they were there. I was pulled into a work boat and rushed to the hospital. Someone handed me my boot. It was still tied. I was bruised from my right knee to my armpit, half my body was a sickening, dark greenish-purple bruise for months, but I was alive. Once again I couldn't walk without limping. I had always doubted people with back injuries; I have since learned to live with mine.

That night, while hospitalized, I began receiving calls from an insurance adjuster for the Cleveland Wrecking Company. He wanted to stop by and have me sign some papers while things were fresh in my mind. I first said no, because of my condition, but he kept calling until I finally told him I had a lawyer, which was not true. The next day I called John McGuinn, local labor and civil rights attorney. Months later I would file a lawsuit against the Cleveland Wrecking Company for their negligence in trying to speed up the demolition which resulted in the separation of the wall from the building.

Jim McNamee had gone home in shock and was taken to the hospital by his wife Linn. When they came to see me, Linn took my soaking wet clothes home to wash. In the right pocket of my overalls was a Timex wristwatch. After washing and drying them she returned them and the Timex watch was still ticking. I always wanted to do a commercial for Timex. It "takes a licking and keeps on ticking" and so did I.

After another nine months out of work, again on "light duty". I could never get a doctor to understand that there was no such thing in my line of work. I walked like an old man with pain in both legs. When I returned to work I kept re-injuring my back and returning to worker's compensation. I felt as though I was caught in a nightmare in which I would starve to death or cripple myself permanently.

I was hired to work for Jensen and Reynolds at the sewer plant on Evans Street in San Francisco. This was the same company I had worked for in Lovelock, Nevada. One day as I was stepping down from the tracks of the crane I caught my foot in the cutting torch hoses I had been using. I fell about four feet face first onto a cushion block hitting my chest just before my face hit the ground. Without the injured back I believe I would have been fine, but the fall put me back on worker's compensation.

For the third time, I was released on light duty and I called the Jenson and Reynolds superintendent Bill Murphy to see if I could come back to work. His answer, "You'll never work for Jenson and Reynolds again. We don't need no more workman's comp claims."

Unlike many pilebutts, I took Emerson Frank's advice to heart. If the union had a job that was only for one day, I would take it. If I had to drive to Nevada for two days of work, I would do it. If everyone was standing around I was working. Still, I was locked in this infuriating cycle. It was all right for me to drive two or three hours a day to work. At times I would only get two days a week when I was healthy. But now because I was injured, this "company man" Bill Murphy would never give me work. And then, some company-appointed doctor who could not understand how physically difficult pile driving was kept sending me back to work to re-injure myself. Additionally, I was sick of insurance men sneaking around trying to get me to sign releases while I was drugged.

Chapter Five: *Finding my Brain*

I was about as far down in the dumps as I could get. My leg had gotten stronger but my back would need to heal for me to return to the work that I loved. I felt as lonely and powerless as I had ever been.

I would not move back into my father's house. He had worked for West Transportation as a diesel truck mechanic for thirty years, but they closed their doors and went out of business when the owner retired. My dad was the last mechanic in the shop, and when he turned out the lights for the last time, that was it. He did find work in the late 70s and early 80s, but things were tough. I couldn't in a million years move back with my mother and father; they still had my little sister Wendy at home and the shame would have been too great.

During this time, I had become an active member in Local 34, attending all meetings of the executive board and the general membership meetings. Having always been a History buff, I was good at asking questions about how and why things were, many of which people could not answer. Finally, after another round of, "what about this and what about that", Local 34 President Gary Bakke appointed me "Historian". I don't think he thought I would take the job seriously, but I went to work two or three days a week in the union hall, emptying cabinets, going through boxes, filing pictures, reading old minute books; organizing everything into one set of lockers the local had set aside for my project. I began to believe again that I was a human being, not some walking mass of flesh…a man that had some value to society, making a small contribution to something bigger than himself.

One night Paul Lewis, the union's Recording Secretary and one of the local's most respected divers, told me that he was going to retire, and asked me if I would consider running for his position. All I would have to do was take notes at the meetings, and the union's secretary would type them up. I was honored and accepted. With Paul Lewis' support, I was elected on a white ballot, which means I ran unopposed.

I started reading labor history at the UC Berkeley Bancroft Library, and began learning the history of our trade, things no modern pile drivers knew. I found myself studying archival techniques, and bought a camera with copy lenses in order to photograph every picture in the union collection. I interviewed and recorded old members such as Jack Wagner, Al Harris, and Ed Koski; once famous but now forgotten.

Although trade unions often cooperate and work together on jobsites, I began to realize why the Pile Drivers Union was so wary of the United Brotherhood of Carpenters (UBC). Local 34 was an affiliated local union, but worked under a separate agreement. I learned why Pile Drivers Local 34 participated in the Carpenters organization in a way that they would be left alone by them. I found incidents in the union's history where the Carpenters district council and international union had tried unsuccessfully to take over the pile driving jurisdiction.

Pile drivers in San Francisco had been organized as an independent local union before international unions came into existence. Local 34 received its charter from the UBC in 1920 only after being forced out of the Iron Workers International Union. The two international unions had bitter fights over the work jurisdictions, which ultimately came to rest under the carpenter's banner.

Local 34 was a progressive union in the 1930s, when the membership voted to support the CIO in their efforts to organize the lumber and saw mill workers in the Pacific Northwest. These same sawmill workers had been members of the United Brotherhood of Carpenters, but marched out of the general convention in protest of their mistreatment by the UBC.

Local 34 voted to send five cents per head to help the sawmill workers organize. The UBC retaliated by pulling Local 34's charter. The membership of Local 34, in defiance of the internationals action, took over the building and the operation of the local union and issued "white cards" to local 34 supporters.

At one point Joe O'Sullivan, the business agent for Carpenters Local 22, went out on one of the docks and claimed the pile driving work for Local 22. Five minutes later he gave it back, and retreated with his brother carpenters to their union hall bloodied and bruised. That may have been the only defeat Joe O'Sullivan ever suffered.

Local 34 and the carpenter's international made peace, agreeing that Local 34 would stay out of the international union's business and the international would leave Local 34 alone. This truce existed for thirty years, until 1997.

I continued to work as a pile driver, and my limp became more pronounced and obvious. Inevitably, I would re-injure myself and return to workers' compensation.

One day I was reading the want ads in the paper and saw a job for a "pile driving hammer maintenance man" with the L. B. Foster Company. The job was maintenance work on rental pile driving equipment. I applied for the job and went to work for $6.00 per hour, half of what I had been making as a journeyman pile driver. My job was menial, service the hammers and repair them in the yard. But at least I could get off of workers' compensation and heal up, and return to pile driving.

Out of every bad thing that has happened to me, something good resulted from it. I couldn't play softball while recuperating, so instead I was asked to coach a woman's softball team. That's where I met Leslie Fouche. I was twenty-eight and she was twenty-two. I initially liked her very much; she was a great softball player and all around athlete. She is the nicest woman I have ever met, and I think I saw her every day until we were married.

It's funny to me that she was born and raised in the same neighborhood as all my friends from Davis Street in San Leandro, though much younger. We went to the same high school, but never met. The first time I noticed her we were at a party thrown by the softball team. She asked me what I did and I told her I was a pile driver. She said, "Really, my grandfather was a pilebutt..." using the same term we do to describe ourselves. She said her father and brother were both union carpenters. My being a construction worker was to her a normal thing, something honorable.

Finally in 1980, the lawsuit I had filed involving my injury at Pier 39 came to court. During the pre-trial settlement conferences, I participated in discussions to settle the case without going to trial.

The lawyer for the insurance company suggested that we go outside; he wanted to make me an offer. This is the honest-to-God truth--he was wearing a polyester suit, a bright yellow jacket with white pants, white shoes and a red tie. He held up a cashier's check made out to me for $75,000. While the largest sum of money I had ever put together was $5,000 I had done the math on how much I had lost being off work, how much I would have to reimburse the insurance company for my medical expenses, and what I stood to lose in the future if I was not able to return to work.

My number came to $100,000, and here was this bald-headed, polyester-suited, insurance company lawyer, sneering. I looked towards John McGuinn, my attorney and one of the best labor attorneys in the United States, and asked him for help. He said, "Mike, $75,000 is a lot of money, and I think we could get a little more if we go to trial, but think hard."

The Cleveland Wrecking Company's lawyer looked at me while holding the check at both ends. He started snapping it in my face and said, "Some one like you may never see this much money again." I thought to myself, what the fuck does he mean, "Someone like me"? I got pissed; I knew I had done nothing wrong, I knew the Wrecking Company had been negligent. I weighed the pros and cons, and I bet on the events that led to my accident holding up before a jury and the ability of my lawyer John McGuinn. What kind of man would I be, and what kind of pilebutt would I be if I let this bald-headed sleaze bag intimidate me. I'd been crushed, almost flattened like a pancake, in continuous pain, making every effort to work, and thrown away like and old pair of socks. Fuck them, I thought, and we went to court.

The trial lasted four days. The lawyer for Cleveland Wrecking was one of the partners in Pillsbury, Madison and Sutro one of San Francisco's most powerful corporate law firms. My first thought was, oh shit, you've done it now, Mike. However, he put the jury to sleep, boring them with mundane information, while McGuinn dazzled them with horrible descriptions of inept supervisors rushing to save a few dimes. The company doctor who examined me described my injuries as a minor bruise. McGuinn produced blown up pictures of the fallen walls, debris floating in the water, and a picture of my bruised and swollen body.

The jury went out to deliberate; we went home for the night. This was the longest day of my life. The next day they returned with a verdict. As I sat waiting, the judge called forward the two attorneys when they returned to their seats. The Wrecking Company attorney leaned over and said to me, "Enjoy your money son, good luck to you, I told them to settle with you".

The judge then read the verdict. I was awarded $275,000, an unbelievable amount. Needless to say, I was in shock. Some of the jurors were crying; one of them, unknown to anyone, was a janitor for the post office who had survived being blown off an aircraft carrier while in the Navy. Others lived with people who suffered from back pain. They had wanted to give me $1 million, but there was an insurance adjuster on the jury who talked them down to $275,000.

I left the San Francisco courthouse on cloud nine. I called Leslie at work, then my mother, and then I got drunk. I quit L.B. Foster and decided to take six months off work.

Chapter Six: *Every Cloud has a Silver Lining*

I went out looking for a house to buy, nothing special, but something Leslie and I could make a start in. Before the trial I knew I had nothing to offer her. At times I thought, what does she see in me? I'm broke down, just getting by. Now, I was feeling pretty lucky. I could buy a house , heal up, and go back to work. And for fun I could go to the University of San Francisco and take Labor Studies courses.

I could move out of the apartment I had lived in for several years, up on "Okie Hill" in San Leandro. It wasn't the finest establishment but my landlord Henry Aubie never raised the rent because I always paid on time. Sometimes, if he had a rough tenant, I would collect the rent for him. Or I would paint for him or work with him handling drill augers when there was no pile driver work. He was a tough old guy but treated me well and I will always remember his rugged kindness.

One day I was reading the real estate section in the local paper and saw a house for sale. It was listed "as is". The owner had moved to Arizona and the house had been on the market for some time. It seemed there was termite damage. One night I went to the house and punched a few holes in the stucco walls to see the extent of the damage, which didn't seem to have gotten that far. I took the so-called termites from the basement to an exterminator and he told me they were not termites, but some kind of moth that eats cellulose (a cardboard box).

I thought that I could fix the house if I got it for the right price. I went to the realtor and offered them $50,000 for a house that was on the market cheap at $75,000. She laughed at me but was bound by law to make the offer. To her amazement and mine the owners accepted it, with the caveat that I buy the house within fifteen days. I wrote them a check on the spot. I tell this story because it was here that I learned a valuable negotiating technique; never accept the first offer and know as much about the issues as possible before you go into a negotiation. You never know what the other side will really settle for.

I spent the next six months fixing up the house at my own pace, and thinking about my life as though I finally had a future. I proposed to Leslie and we set a date to get married. My parents were ecstatic because they loved Leslie. Ironically, both accidents had changed my life for the good. When I crushed my leg, I completed my education and received my Associates of Arts Degree and found that I could coach. Following the accident on Pier 39, I met Leslie, my wife of thirty years. What other good things could come to pass as a result of that wall falling on me? Well, for a start, in 1980, I returned to work as a pile driver. Leslie's father, Roy Fouche, was a business agent for Carpenters Local Union 1622. Her grandfather, Ben Fouche, had retired from Local 34, having worked as a pile driver since 1917. Her brother Robert was a union carpenter. So I fit right in, and her father always respected my role in Local 34 as the youngest guy on the executive board.

Bud Bryant, the Eighth District board member for the United Brotherhood of Carpenters had started pushing organizing by the Carpenters Unions, in the late 1970s. In 1978 the delegates of the Bay Counties District Council voted to create the dues check-off system. Ten cents per hour was negotiated into the collective bargaining agreement to hire union organizers. By 1980 the Bay Counties District Council of Carpenters was putting an organizing program together. And in 1981 the Eighth District and the Bay Counties put together an organizing drive. That first team was Babe Garcia, Fred Calloway, Bill Sweeney, Ken Viscovich and one more fellow I can't remember.

One afternoon he invited me to San Mateo County with a group of union organizers. The organizers would go out into the field in teams. As they were one person short, they invited me along.

I spent the next six months organizing in San Mateo and Sacramento. I never received a paycheck during that time and liked serving as a picketer for Local 34. I felt it was my duty to pay something back. It was there that I met Babe Garcia, a member of Carpenters Local 586 and a house framer. He was no different than the guys I had grown up with and as tough as any pile driver in Local 34.

One day we were riding together and I asked him where he was from and he said Vacaville, California. I told him that my mother's family was from there and that I had many cousins still living in the area. We discovered to our amazement that many of my cousins were his cousins. We have been cousins ever since, no matter how distant.

My great uncle Blas Sanchez had told me about the strike in Vacaville by Spanish Tree Pruners in 1932. My mother's family had moved to Oakland following the strike and my uncle Blas had been black listed as a result of his participation. Babes step-father Henry Espinosa had been arrested and held in jail while other union organizers were kidnapped from the Vacaville Jail, horse whipped, stripped naked and painted red.

Babe Garcia is one of my dearest friends and someone I have emulated as an organizer for his toughness, intelligence and his integrity. From the first days, he taught me that working for the union was a privilege very few people ever enjoy. That because we were hired, we needed to work harder, that hard work was the only way the union would change for the better. And the only way the members would ever respect organizers.

Babe had the greatest nose for bullshit, and would be the first person to call anyone on it. It was Babe Garcia who taught me that confrontation of the boss in front of the workers was a powerful tactic.

He also did not couch his words, even if you were the General President of the UBC. After we lost at the 1991 convention, Babe walked up to General President Lucassen and told him, "You won't have to fire me, because I won't work for you." He is one of the most honest men I have ever met and the person who taught me that love for one's brother is a primary principle of trade unionism and union representation.

In 1985, during the Louisiana Pacific strike, the National Lumber and Sawmill Industry was holding its national convention in San Francisco. Babe and I made a large sign that said "Boycott LP". We secretly made our way into the auditorium of the St. Francis Hotel and hung the banner behind the curtains on the stage where the podium was located. The guest speaker was the General Council for Louisiana Pacific and one of Ronald Reagan's appointees to the Department of Agriculture. As he was introduced, I raised the lighting behind the curtains and then opened them. The audience erupted in laughter and we ran like hell out the back door.

Jim Green – Courtesy of the Northern California Carpenters Regional Council

Chapter Seven: *Jim Green*

In 1978 there were nine carpenters district councils throughout Northern California. There was one independent local union that did not belong to any district council, there was no central authority that oversaw northern California carpenters. The Master Agreements were the only rules on which they would all agree, but each district council informally amended the Master Agreement that carpenters worked under, in each area to fit the desires and enforcement of the local business agent.

The head of the Bay Counties District Council of Carpenters was a man named John Watt who was not a carpenter by trade. He had been the City Manager of Belmont, California and had found his way into the position of Secretary Treasurer.

His authority was limited and his political powers lay in the hands of the elected business agents from the various local unions. The position of Executive Secretary Treasurer had been held for years by Abe Muir, Dave Ryan, Chet Bartalini, Al Figone had always been held by a union member. At the time the elected senior business agent had some authority over the elected agents in their local union. But they could not hire, fire, or discipline any business agent nor could the executive board of the local union. Factions could campaign to elect someone new during the next cycle, but for all intents and purposes an elected business agent could either work at their job or not work during their elected term in office, and there was not much anyone could do about it. The District Council Secretary was a figure head with limited powers.

John Watt convinced the district council delegates to vote on a dues check-off system. Dues check-off meant that the members could voluntarily sign a card which allowed the trust fund to direct ten cents per hour into a fund which paid for the union to hire organizers. The delegates voted to establish the dues check-off contribution and build up enough money to hire organizers.

The fund quickly grew and in 1981 the first organizers were hired--Roy Fouche, Bill Sweeney, Gary Ross, Babe Garcia and myself. As Director of Organizing, Roy Fouche was the first director to work independently and separately of any local union. It was then that I met one of the most brilliant trade unionists I have ever known...the person who laid the foundation for a union where union carpenters could work for one employer all year and have their benefits follow them throughout Northern California. That man was Jim Green.

In 1980, Jim Green left his job as business agent of Millwrights Local 102 to work as John Watt's assistant at the Bay Counties District Council.

Jim Green was perhaps the most successful organizer in the Bay Counties District Council of Carpenters. He was the son of Ray Green, the former senior business agent of Millwrights local 102. Like Chester Bartalini, Ray Green was an old-style rank and file trade unionist.

Chester Bartalini was a member of Carpenters Local 36 in Oakland, California. Bartalini worked in the shipyards and was a strong advocate and a New Deal Democrat.

He had moved to California from the coal mines of Indiana. Because of his strong support for unionism, he was forced to leave the United Mine Workers in fear for his life and his family's after a long and bitter strike.

In 1953 and 1956, Bartalini was the Executive Secretary of the Bay Counties District Council of Carpenters. He led the fight for the Carpenters' Health and Welfare Plan and the Carpenters Pension Plan, and is considered by many to have been one of the strongest labor leaders of his time.

If there is a genealogical line of political affiliations within a union, then Jim Green is a descendant of Chester Bartalini and so am I. Jim Green was raised hearing stories of trade unionism and the hardships of the 1930s. His father and mother had been cotton pickers in the fields of Texas and by their own admission thought of themselves as Okies, a sometimes derogatory name for Midwestern agricultural workers and farmers who migrated to California during the Great Depression.

Jim served his apprenticeship and quickly moved through the chairs of the local union executive board. By 1976, he had been elected senior business agent of Millwrights Local 102.

Having seen leadership ability in Jim Green, John Watt and many other progressive trade unionists backed his candidacy for Executive Secretary Treasurer of the Bay Counties District Council.

Within two years, Jim convinced the elected business agents and the district council delegates to amend the bylaws of the Bay Counties District Council to support the hiring of business agents. This was a radical change and departure from the existing process, but because the old structure had many flaws and the argument for accountability of the leadership was sound, a majority of rank and file delegates were convinced to support this radical change.

Jim Green's political abilities and negotiating skills would mold the Bay Counties District Council into one of the most powerful organizations in the UBC.

The act of restructuring unions has been around since the founding of the UBC. Carpenters local unions were active in the San Francisco Bay area long before the formation of the United Brotherhood of Carpenters in the 1800s.

In general the structure of the UBC went from the bottom-up. The members belong to a local union. The local unions were chartered by the International Union and belonged to a District Council. The District Council was also affiliated with the International Union. Now days the District Councils have been consolidated into Regional Councils which absorbed the District Councils into multi state Regional Councils.

Since the founding of the UBC there have been hundreds of local unions merged, consolidated, disbanded or who have just closed their doors. The arguments that took place during these activities created dissention and controversy. Consolidations and mergers of local unions tend to be the most disruptive actions a union can take. Even if the numbers of members that local unions hold are significant, this action can cause friendships to end, fights to break out and members to leave.

From my conversations with Jim, I gathered that his main reason and concern for merging the local unions in the Bay Counties District Council of Carpenters was to eliminate costly redundancies and to consolidate the Bay Counties District Council into a bigger and more manageable organization.

The Master Agreement or union contract was interpreted differently in each local union jurisdiction. For example, San Francisco, a city of eleven square miles, had three local unions. Local 22, Local 483 and local 2164, in addition to Pile Drivers Local 34, Millwrights Local 102, Shipwrights Local 1149, Cabinet Makers Local 42 and Industrial Carpenters Local 2565. Each one had their bylaws and rules to enforce. They also interpreted their agreements differently than the carpenters interpreted their master agreement.

Each local union had factions that fought for control of the local union and each local union fought with the neighboring local unions in the city or county over territory and hiring rights.

In San Francisco, which was divided into three areas, locals would rotate every three months. Areas would change and a different local union would replace and police the area. As the areas changed, so did the dispatching of work to the members; the local unions would replace the stewards on every job to make sure that only carpenters from their local would be dispatched out on the job site. This ran afoul of the out-of-work lists, originally formed to guarantee fair awarding of work to the membership. It is easy to see that this system was wrought with inequities.

In another example, Contra Costa County had two local carpenters unions, 2046 and 642. At election time at Local 2046 in Martinez California, the senior business agent Frank Castiglione would make a few calls to his contractor friends and the whole crew would be ordered to put down their tools, get into company trucks, drive down to the local. At the local union they would be given a slate card and then instructed on whom to vote for. They would then vote, return the slate cards and return to work, never missing an hour's pay. Meanwhile, any person running in opposition would have to take time off from work, campaign for every vote and spend money on political mailings. Castiglione and his slate were unbeatable and anyone who ran against him might have to move their union book so that they could find work in the future.

Alameda County had Locals 1622, 194 and Local 36. They fought with one another in many of the same ways as the other areas. The situation was chaotic and disorganized. As organizers, what was permissible in one area was not in another. Some contractors were untouchable and secret agreements between business agents and contractors were everywhere.

Then, throw into the mix the same unwillingness to cooperate between the nine or so Carpenter District Councils in Northern California. On more than one occasion the Bay Counties District Council of Carpenters targeted an employer for antiunion policies. Neither the Santa Clara District Council of Carpenters nor the Sacramento District Council of Carpenters would cooperate in the organizing effort, leaving the targeted employer free to work nonunion in other district councils. This made it more difficult to create enough pressure on the employer for him to make the move to the union side of the street.

So, overall, what we had were three or more independent local unions, thirteen independent district councils, and three Master Labor Agreements dividing the State of California. The Northern 46 Counties Master Agreement and the Southern 11 Counties Master Agreement divided the State of California into two more separate bodies which acted independently of one another.

Local unions fighting one another for control of work and dispatching, neighboring counties fighting each other for work and dispatching, and a state divided in half fighting each other for work dispatching and trust fund contributions.

Meanwhile, a growing nonunion sector was going about its business without disruption because the local unions, district councils, and agents were too busy taking work opportunities from union members from various other local unions and ignoring the nonunion contractors who were slowly taking union jobs.

Worst of all, union agents became satisfied with the number of members they had. As long as the membership was employed and the agent could be re-elected there was no need to build the union. Nonunion workers who approached the union for help in organizing were ignored.

This disorganization and building of small kingdoms grew even farther out of proportion when the other trades became involved and the fights over jurisdiction began to spin out of control.

This scenario still existed in the rest of the building trades unions. Electricians and Plumbers in San Francisco could not work under the terms of the San Francisco agreement across the Bay in Oakland. Nor can the employer take his crew from San Francisco to Oakland, eleven miles away. The employer had to lay off his crew and hire a new crew from the Oakland local union, cutting into his ability to maintain a crew of qualified journeymen. The employer could not make contributions on behalf of his men once they leave the boundaries of their local and trust funds, even though they were all union members. As a related aside, the carpenters had the foresight in 1957 to create one Master Agreement and trust fund from which the members' benefits could be collected and distributed. Even today many building trades unions have separate trust funds within a small geographic area, each administered by separate boards of trustees and each negotiating for a more expensive benefit for their members.

By 1978 membership had declined to an all-time low. Employers were frustrated by the inability to guarantee union members work on union jobs. They couldn't keep good crews working if their jobs left one local union's area and moved to another. They were required to lay off a member from one local union and hire an unknown carpenter to work if the job was in another area, and heaven forbid they take a crew into another district council. These disputes and gamesmanship were ripe for union busting and in 1982 we got it. Ronald Regan was elected and the war against the Labor Movement began.

In reality, the ability to hire organizers was due to the monthly dues check-off. The ten cents generated by the dues check-off, multiplied by one million hours per month created a substantial amount of money to the local unions. I still remember when Pile Drivers Local 34 had to sell the business agent's cars so that they could make payroll during the 1970s. It is dues check-off that has made the carpenters union the viable and vibrant organization that it is today.

Jim Green convinced small local unions to merge into larger local unions, one per county. The restructuring was not easy; there was tremendous apprehension and mistrust by many members because they feared leaving too much power in the hands of one person.

These mergers had to be done in a legal manner and approved by the General President of the UBC, Sigurd Lucassen, who was merging other district councils himself. It was obvious to Jim Green and the executive board of the Bay Counties District Council that Northern California would not be left alone.

Green told the elected business agents that the Bay Counties could conduct our own mergers or wait for the General President to come in and do it for us. This second option was much less palatable and the mergers, while resisted, took place. The first step was to get the General President to allow the Bay Counties to do this ourselves.

So, on April 7, 1988, General President Lucassen restructured the Bay Counties based on the recommendations made by the bay Counties Delegates and interim committees were formed to facilitate these consolidations as follows:

Millmans Local 42, Shipwrights Local 1149 and Furniture Workers Local 3141 formed a new Industrial Union 2236. In San Francisco, Carpenters Local 22, Local 483, and Local 2164 were consolidated in Carpenters Local 222, which would some time later reissue the Local 22 charter. In Alameda County Carpenters Local 36 and Local 1622 were consolidated into Carpenters Local 713. In San Mateo, Carpenters Local 162, Local 848, and Local 1408 were consolidated into Carpenters Local 217. In Contra Costa County Carpenters Local 642 and Local 2046 were consolidated in Carpenters Local 152. And Carpenters Local 180 in Vallejo was ordered to affiliate with the Bay Counties District Council.

All the funds and assets of the consolidated local unions were ordered transferred to the newly formed local unions. New elections were ordered to replace the interim executive boards formed to facilitate the mergers. The election for the Bay Counties District Council was delayed until new delegates could be elected by the newly formed unions and all the newly formed unions were ordered to affiliate with the Bay Counties District Council of Carpenters.

In San Francisco and San Mateo, the old guard of business agents were so upset by all the change, they convinced their executive boards to spend the assets of the local union before the mergers took place. This is best illustrated by what occurred in San Francisco, where Local 22 had long been aligned with the San Francisco Building Trades Council and Local 483 chose to participate in the San Francisco Labor Council. Both had longstanding feuds dating back to the early 1900s. Local 22 feared what they thought were the socialist tendencies of Local 483. The Labor Council represented unions not considered to be tradesmen but rather industrial and public employees who tended to be more liberal and representative of lower paid union members, while the Building Trades Council saw themselves as equals of professionals, like doctors and lawyers. Compounding this was an active and independent rank and file movement within the carpenters union, which held organized meetings as far back as 1926. However, the most controversial result of the consolidations was the vote to hire business agents who had been elected in the past, and the decision that union hiring halls would be controlled by the district council. On July 26, 1988, Local 22 filed for a restraining order to stop the consolidations, but the effort was unsuccessful.

As a sign of their displeasure, Local 22's Executive Board held a huge dinner and bought jackets for every member leaving the local broke at the time of the merger. Today they rent a building owned by Local 2236 and have gone with hat in hand to the regional council to raise enough capital to build a union hall. Had they not wasted the union's treasury they would have a fine building somewhere in San Francisco today.

Green reasoned that hiring rather than electing business agents was justified. Elected agents had created little kingdoms which excluded people from union membership, and members who would not cooperate were often forced to move their union book to another local union and were unrepresented or unemployed. The hiring halls were manipulated by the agents to reward their supporters and punish those who might run against them.

The hiring of agents and control of the hiring halls brought accountability to a higher authority. Business Agents could no longer refuse to cooperate in the best interests of the membership and would be held accountable for their actions to the contrary.

The International Brotherhood of Carpenters watched and then began to merge district councils prior to the 1991 General Convention. It was at that time that the old North Coast District Council of Carpenters, Locals 751, Ukiah and the Eureka were merged into the Bay Counties District Council of Carpenters.

In fear of being merged into the Bay Counties District Council, the remaining smaller district councils were merging into the Sacramento District Council of Carpenters and the newly formed Central California District Council of Carpenters.

At the end of the process, the number of district councils was reduced from nine to three. From 1982 until 1997, the carpenter's master agreement was negotiated by Jim Green, and eighty-five percent of the work was organized and performed in the Bay Counties District Council and the Santa Clara County region.

Also, beginning in the 1980s with the Reagan administration, working people faced the most dramatic changes in working conditions and economic strife since the Great Depression, a situation that continues to worsen today. Neoconservative, union-busting, free-market policies that pander to Corporate America and the rich have decimated the American working class, and are steamrolling through the middle class as well. Back then, Jim Green was required to make decisions to protect the best interests of the membership and preserve the carpenters union. During Green's tenure, working carpenters faced Ronald Reagan, Georgia Deukmejian, George Bush I, and Pete Wilson. Each Republican administration's policies created a recession which forced union members into the unemployment lines, and following each recession we grew by contractors and man hours.

The dramatic changes in the structure of the District Councils and the formidable political climate of the 1980s and 90s made Jim Green one of the most important men in the California labor movement. Jim Green became part of the fight to change the United Brotherhood of Carpenters and Joiners. He stood up against the $100,000,000 loss of Brotherhood money. Unfortunately, his support for Whitey Rogers for General President of the United Brotherhood of Carpenters and Joiners of America ultimately cost him his job. It is a fact that during the 1991 convention, both sides vying for control of the United Brotherhood had asked Green to be on their tickets. Jim's political aspirations ended as the head of the Bay Counties District Council of Carpenters. Had he desired the top spot in the United Brotherhood of Carpenters, he would've only needed to make the appropriate political decision, and the general presidency might have been his rather than Douglas McCarron.

Because of Jim Green, a union organizer can follow an employer into any area in Northern California. Past organizing campaigns were limited to the boundaries of the district councils, and on many occasions, were stopped dead in their tracks when the targeted employer crossed from one county into another, or district leadership changed. For example, during the campaign against Seeno Construction, the employer continued to work unmolested in the Sacramento District Council region. During this time, we were conducting a joint organizing campaign with the Sacramento District Council. As the campaign progressed the Executive Secretary of the Sacramento District Council, Jim Larsen, was voted out of office. The new Secretary, Leon Wagner, was not committed to organizing, and within two weeks our joint organizing campaign died from lack of support. It took two years for us to organize the employer. Now, an organizing campaign is no longer limited to the geographic boundaries within northern California.

Jim Green's vision built the foundation of the Northern California Carpenters Regional Council. His ability to build by consensus rather than mandate made the consolidations in Northern California possible and palatable to a strong and vocal rank and file.

It is because of Jim Green that Northern California is the third largest regional council in the United Brotherhood of Carpenters. Though geographically it is the smallest regional council in the Brotherhood, his vision and abilities to move political agendas through the local unions and to the highest echelons of international union politics made him one of the strongest leaders of the United Brotherhood of Carpenters in modern times.

The model Jim Green created in Northern California was different from the model used by Douglas McCarron to restructure the United Brotherhood of Carpenters. I personally supported our method because it created an environment conducive to organizing. Our model allowed participation by the local union membership and delegates.

I regret that some democratic freedoms have been lost by union members. Having seen the corrupt practices by some elected business agents, I believe that only accountability and the possible loss of one's job when an agent is found to be lacking in ability or corrupt and derelict in his duties, is the solution. Additionally, I do believe that the person at the top, the Executive Secretary or General President, should be elected by every dues paying member, and not by delegates.

Chapter Eight: *Union Organizer*

Sometime in the middle of 1981, the Bay Counties District Council of Carpenters put out a call to all affiliated locals. They were accepting resumes for Union Organizer positions. This was the culmination of decisions made as early as 1978 to start organizing to meet the declining numbers in the local unions. I felt qualified for the job because I was an elected officer in my local, I had worked on the organizing drives in San Mateo and Sacramento, and I had signed nonunion employers to the Master Agreement while working voluntarily. But the truth is that my wife's father Roy Fouche was the Director of Organizing. So, I may have been qualified, but I was hired out of nepotism.

In November of 1981, I started my new job. Roy Fouche told me that under no conditions would I have it easy. In fact, I would have to work harder than the others or he would fire me. I have to say I was as happy as I had ever been. To my knowledge this was the first Construction Organizing Department in the United Brotherhood of Carpenters.

Roy Fouche, Bill Sweeney, Gary Ross, Babe Garcia and I were hired and we worked out of the BCDC Headquarters at 995 Market Street in San Francisco. So like I had done as a pile driver I put my head down and went to work. From that day forward a wedge was driven between me and my old friend at Local 34, Terry Price.

Neither hard work nor signed contracts would make up for the fact that I worked for the carpenters and the boss was my father in-law, something that others whispered behind my back, but never said to my face.

My first year was spent driving from job to job filling out a job call report, checking subcontractors for union affiliation and checking union cards of the men on the job. We would find nonunion subcontractors working for union general contractors, and nonunion men working for the same subs. The master union agreement requires all subcontractors working for union contractors to be union also. When this happened, the nonunion subcontractor would sign the agreement or leave the job.

Remember, organizing was a new concept to the Bay Counties District Council. Job sites were policed or not policed, and the subcontractor's clause was enforced or not enforced. The first years of the department were spent on union jobs signing nonunion subcontractors who had been allowed to work in violation of the master agreement.

Another very popular business venture by the union contractors was to operate a nonunion company. This practice, known as double breasting, was hard to prove at the time because so many union companies also had nonunion operations. I spent lots of time learning to use the computers at the District Council to expose double breasting contractors. Through the auditing and grievance process agreed to in the master agreement, good fieldwork and good research, union employers were forced to put these bad companies out of business.

In today's world this is considered contract enforcement and is done as a matter of routine by business agents. But the system was different then - business agents I was working with were elected by the membership and they did what they pleased. If a carpenter pissed them off he didn't work. If he was not white he was possibly shuffled off to another local. If he didn't vote for the "right" business agent he didn't work.

For some BA's, the day ended at the bar at noon. There were jobsites we were told we could not go on, and when we did we would always find nonunion men and subcontractors. There was no accountability to the members. These irresponsible agents, elected by such a small minority of carpenters more interested in working than in the good of the whole membership, often abused their positions.

These few irresponsible agents left a legacy of untrustworthiness that the employers could use to exploit workers and divide the union from the membership. I must say that most were honest, hard working, dedicated business agents. As I write this chapter thirty years later, Bob Alvarado, hired as an organizer is now an Executive Secretary Treasurer of Northern California, and ninety percent of the current field agents were organizers. The Bay Counties District Council was light years ahead of the rest of the labor movement.

I was doing well, but I wanted to organize more so I read every union book I could get my hands on. I also began reading about military strategies, and biographies of leaders from Gandhi to Clausewitz. We were learning about confrontation and direct action, working while others were at home, not reinventing organizing, but applying organizing techniques successfully used by others in the past.

One day in 1983, I was working alone in the Financial District of San Francisco, checking union cards to see if there were any nonunion men working in a new building on Stuart Street. I asked this guy Manuel Figaredo for his union card. He was moving Modular Office Systems into the building. Manuel handed me a union card from the Plastics and Novelty Workers Union that had expired three years earlier. I checked more workers and was handed expired union cards from other unions.

I found out from Manuel that he worked for the Rucker Fuller Company, and that he wanted to be in a union. He was tired of being second class and having to hide whenever the union representatives came into a building he was working in. He told me his boss's name was Bernie Sweeney and that if I needed to talk to someone Bernie was the godfather of Office Modular Systems in San Francisco. So I went to meet him.

He was a very likeable working class guy. He said, "I want my guys in the union, they want to be in the union. But I compete with furniture moving companies, not construction companies. If I were to pay journeyman carpenter rates, I would never win another contract."

I found that he controlled a major portion, and influenced even more still, of the downtown San Francisco modular furniture work. I began raising hell, picketing jobs downtown, and showing up in front of buildings at night. By talking to all the men installing furniture systems, I identified the rest of the nonunion installation companies. I learned that involving the nonunion workers in the effort gave me information and power over the boss. I found out that the building owners and managers feared any kind of disruptions on their jobs, but especially in front of their buildings. And better yet if there was new construction in progress inside a high-rise, the union men would walk off. The financial district of San Francisco became my poaching ground.

I learned that metropolitan areas and high- rise buildings are more susceptible to jobsite action due to the number of businesses they contain, the clientele, workers entering the buildings and the public passing by on a daily basis. The owner could not stand up to the chaos I was able to create.

I had an industry that I believed I could organize, not just one employer who was working as a subcontractor to a union general contractor. My problem was not the employer fighting the union; I had many contractors who wanted access to union projects. The stumbling block was officials inside the union who could not think outside the box enough to even consider any other type of union member other than a journeyman carpenter working under the master labor agreement.

I worked with Roy Fouche, Jim Green and Bernie Sweeney and we negotiated the first Office Modular Systems Agreement. I signed five employers and took fifty applications for membership in one day. That was 1984, over the past twenty years they have reported hours on as many as 500 men a month. Manuel Figaredo, Luis Hernandez and other installers have now retired and collect pensions from the pension plan.

Chapter Nine: *Giacomo Patri*

As I continued to evolve, I was meeting working people who would stand up and fight to protect what they had, people who thought about who they were as citizens and members of the working class. I was developing a conscience and a sense that I was doing something that mattered.

This next step in my development began to crystallize when I found a small booklet in an envelope while cleaning out the old files at Local 34. "Steward on the Job", written by Jack Wagner, was presented in pocket book style, so that a worker could put in his back pocket or his front shirt pocket and carry it with him.

Wagner created the book for union stewards and rank-and-file members, so that they could understand contract enforcement and proper conduct by union members when representing workers and the union. Local 34 is very unique in its steward structure. Each crew can decide among themselves who will be the Steward. They rely upon the pile driver they choose to protect them at the first step of representation.

What struck and eventually changed me were the pictures- -nine linocuts drawn by an artist named Giacomo Patri during the 1940s. When looked at in sequence, these cuts tell the story of the union steward and the enforcement of the grievance procedure. Without reading a printed word, any worker could see the pictures, develop questions, and find answers.

Reading for me has always been difficult. In school I always struggled to comprehend what the words meant. What I see, touch and smell has always had greater influence upon me. An analogy might be in soccer, you see the opposing player, the angles he takes, his confidence, his willingness or reluctance to engage in contact. Or in poker, how someone bets; do they hesitate, how much

they bet, their posture, things I can see or sense visually. Or in baseball, you see the pitcher throw the ball towards the batter and your mind and body anticipate where the ball can be hit.

Patri's illustrations explained to me a concept, something I could not see in my mind when reading written words. These pictures, simple black and whites drawn before I was born, flipped a switch in my mind. At first it was the artwork, then the realization that a book could be simple yet deliver a message. If these drawings existed others must. I became a collector of Giacomo Patri's work.

My use of pictures in organizing stems from these ten linocuts. If simple pictures shown in sequence can make working people begin to ask difficult questions and seek the answers, then they could lead more to action.

Throughout my career as a union organizer I have found pictures, especially cartoons, to be more effective in uniting people against the boss. I have written lengthy leaflets to workers explaining the virtues of unionism and why they should vote for the union, only to find them on the ground or left unread. Then I have handed out pictorial leaflets making fun of the boss or supervisors that were posted everywhere, leaving the bosses to defend themselves against a joke.

My greatest use of visual organizing was against John Casey during the elections for NCCRC Executive Secretary in 2001. While I'm not an artist, my computer allowed me to cut and paste existing cartoons and adapted them to my fight.

The fewer words used, and the more visual an organizer's message, the better the chances are of winning. The hardest thing to teach new organizers is how to listen. Organizing campaigns are not won by writing letters to workers. Direct contact and education leads to ownership of the organizing campaign and then action by each worker.

During one campaign, I was the only organizer, trying to win an election at LTD Ceramics, a 170-employee machine shop. Just days before the election, the company planned a big party in appreciation for everyone's hard work, but I knew it was designed to force people to publicly support the company and overwhelm those too timid to openly support the union.

I decided that since I could not stop the company from having a party, I created a leaflet that looked like an invitation to a going away party for the plant manager. I made fun of the manger and supervisors; the effect was so great that the workers were openly laughing at supervisors. The leaflet was posted everywhere in the plant and the company had to spend time looking for the people responsible for distributing them, and I won the election.

In another effort at Microphor in Willits, California, the company made self-contained toilet systems for passenger trains. As in many organizing campaigns, the company holds more and more meetings with the workers on company time to convince, threaten, and alienate workers.

After one of these captive audience meetings, I was stationed outside the building with thirty rolls of toilet paper all individually wrapped in paper which

had a line drawing of a man farting; on them I had written statements ridiculing the boss and promises he had made to workers if they voted the union out. They were stacked on my car as though on display in a grocery store.

As the workers left the plant I held a union meeting outside on the public street. The workers all took a roll of toilet paper and were laughing as the boss came out in an effort to win support by confronting me. The workers began playing touch football using the toilet paper roll as the football. As he was explaining that making fun of the company's main product was another example of how uncaring I was about the company and its employees, one of the workers accidentally hit him in the head with a long bomb meant for a coworker who was going deep.

At that point he lost all respect from the workers, and I had defeated him without saying a word. I believe that working people in general are not readers. That doesn't mean they are dumb or incapable of understanding ideological issues. It means their daily lives are consumed with making a living and trying to do the best they can. So messages made brief, to the point, and as visual a possible have always been more effective to me. That's why television is such a powerful medium.

Archie Green Market St – Courtesy Luann Green

Chapter Ten: *Professor Archie Green - Shipwright*

Some time in 1982 I received a phone call from my dear friend Ed Kelly, Financial Secretary of Pile Drivers Local 34. An old guy named Archie Green had come into the office looking for information about the local. Ed told him that he should talk to the local's historian. He took the old man's number and said he would give it to me.

Always one to talk about local 34, I called Archie. It turned out that Archie Green was an author, a folklorist, a PhD, and a journeyman shipwright.

Following his graduation from UC Berkeley in 1939, Archie was accepted into the California Conservation Corps. The CCC, as it was called, was a program of Franklin Roosevelt's New Deal - where young men worked maintaining the national recreation areas.

More than 300,000 young Americans joined the California Conservation Corps, escaping the Great Depression and sending money home to their families. Of those, only two percent had a college education. Archie is the only person I have ever met who volunteered to join the working class.

In February of 1941, after his time in the CCC, Archie became a member of Dry Dock, Marine Waysmen, Stage Riggers & Helpers Local 2116 and remained a member of the United Brotherhood of Carpenters for sixty-seven years until his death in 2009.

It was there that he was partnered up with shipwright Ben Carwardine, a Scotsman and Clydesider. The term refers to radicals all over the west coast of

Scotland, but is often more specifically used to label Scottish ship builders who learned the craft on the docks of the river Clyde outside Glasgow.

According to Sean Burns, these Clydebank shipbuilders were vital to British naval production in World War I, but, reduced production after the war drove many of them to immigrate to the United States. Revered for their engineering skills, "Clydesiders" were equally known for their unique brand of militant industrial unionism and Scottish cultural pride.

Even more, Carwardine was a syndicalist whose ideology proposes a political movement controlled by workers that regulates both industry and government. Taught to him on the job by his mates, it was passed on to Archie as easily as his lessons in measuring or tying a knot.

Imagine the waterfront of San Francisco in 1940. Six years earlier the San Francisco General Strike had taken place. The shape-up had ended, heads had been bashed, and men were shot dead on Mission Street during combat between strikers, scabs and police. Longshoremen, pile drivers, and shipwrights - all of these were workers striking against the bosses. The unions won, the shape-up ended. A new progressive era of labor power returned to San Francisco and Northern California.

Carwardine and other shipwrights molded Archie and stimulated his mind and imagination, instilling in him the pride in craftsmanship and true fraternalism that exists in real trade unionists. Archie had gone native; he had joined the tribe of shipwrights, whose hopes and dreams were of living a decent life and dying with dignity and the respect by their peers?

Archie also worked with Chester Bartalini, who would become Executive Secretary of the Bay Counties District Council of Carpenters. And lead them in two long and bitter strikes for Pension and Health and Welfare Benefits.

In 1941 Archie was elected Recording Secretary of Local 2116 and soon found that he was embroiled in the Local's politics. He became a delegate to the Bay Counties District Council of Carpenters and threw his support behind Chet Bartalini and the New Deal.

In 1942, Archie enlisted in the United States Navy and was shipped out to China working as a shipwright until the war ended. Following his discharge, Archie went to work for the American Veterans Committee, placing veterans returning from WWII into union apprentices. He also taught from 1945 to 1947 at the California Labor School, which was accredited by the California State Department of Education for veterans' education under the G.I. Bill of Rights which by 1947 had 220 full-time students.

In the 1952 minutes of the Hearings by the House on Un-American Activities I found that Archie was called before the commission and questioned about his relationship with the California Labor School.

Archie's radicalism was almost anti-radical in that he believed in the rights of workers to organize, form unions, and fight for a living wage. He was anti-communist, but a supporter of Syndicalism. He was the first non- Spanish person I ever met who understood my family's anti-communist views, stemming from the Spanish Civil War and Stalin's purges of the Spanish anarchists.

Because of his open support for Chester Bartalini and his opposition to the entrenched Republican mentality that pervaded the Shipwrights leadership in the 1950s, he left the shipyards and moved uptown to work as a carpenter until 1958.

The 1950s were turbulent times for carpenters. In 1953, they went on strike for three months to fight for health and welfare benefits, and then again in 1956 for pension benefits. The introduction of a pension plan, health benefits, and vacation pay were violently opposed by the employers, but San Francisco was a union town and men like Archie Green and my father-in-law Roy Fouche stood tall and together in these fights.

In 1958 Archie left the trade and returned to college, receiving his Masters Degree from the University of Illinois where he taught until 1972. In addition he received his PhD in Folklore from the University of Pennsylvania in 1968.

During the 1970 he moved to Washington D.C. and worked for the AFL-CIO's Labor Studies Center and the Smithsonian Institute presenting workers traditions on the National Mall for the American Bicentennial Celebration. It was also during this period that he began to lobby for the creation of the American Folklife Preservation Act. His writings on labor folklore and hillbilly music brought him acclaim throughout North America. Archie retired from the University of Texas in Austin in 1982, returning to his home on Casselli Street in San Francisco.

In retirement, Archie was the backbone of the Bay Area Labor History Workshop, The San Francisco State Labor Archive, and the Fund for Labor Culture. All receive tremendous support from the Labor Councils, Building Trades and Local Unions in Northern California. Archie was the mainstay in all these organizations and a relentless fund raiser for the Fund for Labor Culture and History of which he was the founder, Secretary Treasurer and one-man dynamo.

On that day in 1982, I phoned Archie and made my first pilgrimage to Casseli Street in the Castro district of San Francisco. Our friendship was instantaneous-

-his encouragement and support lead me to many projects, including my book, Pilebutt: A Collection of Stories about Pile Driving. I remodeled the union hall building, a museum, and photo gallery exhibiting our work. With the help of Jim Johanson we built a memorial to the members who had been killed on the job. And with Maria Brooks' great work we produced the movie, Pilebutt: Working Under the Hammer.

Archie and I shared a common genealogy through Chet Bartalini. We were brother members of the Carpenters union, we both worked on the water, and were delegates to the District Council of Carpenters. This union genealogy is mystical to me, in that we were both supporters of the same people who molded the union long before I was born... a genealogy of fraternalism and collective friendship.

After twenty-seven years as a union organizer, I've learned that no union campaign can be won if the organizer does all the work. If workers do not believe, take control, and share the risk in winning, the union campaign is lost. Archie knew this and used these very same skills in recruiting folklorists. Those lucky few would go on in pursuit of their own projects, accomplishing great works that define and redefine folklore, in word, song video, and Internet, and believing that their work was more than a job but a contribution to mankind. No matter how obscure.

The love that exists between men and women who work together everyday, suffering the pain and misery of work, sharing the lives of each other's families, makes of them brothers and sisters. That bond is greater than blood, a loyalty that binds human beings together as individuals, as union brothers, sisters, friends and tribe members.

I know that Archie believed as I do, that, fraternalism, love and respect are the reasons unions exist today. That it is the glue that held us together; love, mutual support, and an understanding of one another without spoken word. This bond is the secret cement that is the foundation of true unionism. Archie learned this from the Scottish Clydesiders as I learned it from Okie and square-head pilebutts. We both learned that brotherhood, friendship and love often unspoken, and at times telepathic - exists in the hardest of occupations in the darkest, most dreadful, and dangerous places.

Chapter Eleven: AETNA *Life and Casualty*

In 1982 Jim Green was elected as Secretary Treasurer of the Bay Counties District Council of Carpenters to replace the retiring John Watts. I will always be grateful to John Watts for giving me a chance to become a union organizer. John had been the mayor of Belmont, California before becoming Secretary Treasurer of the BCDC. I have no idea how this took place, but John Watts and Bud Bryant had the foresight to reinstitute organizing departments long before the UBC or any other building trades union.

In 1982 the Carpenters Master Agreement expired and the carpenters went on strike against the homebuilders. About 65% of tract homes were built union in those days. In hindsight, this was the beginning of the end for union homebuilding in Northern California. After a short strike the commercial contractors had signed the new agreement. But, the home building contractors who had hired carpenters directly had sub contracted all carpenter work to union workers before the strike. Legally we could not picket jobs where everyone was union. The homebuilding contractors then became developers with no carpenter employees.

Under Jim Green's direction we struck the Home Builders. It was quite illegal. The offices began to fill up with lawsuits, subpoenas, and depositions. The stacks of paper were huge and lined the conference room. One day I was sent by Roy Fouche out to a picket line in Hercules, California, where things were getting out of hand, I was sent there to calm things down. It was a Centex Homes job site and the carpenters on this job were nonunion from Los Angeles. Like any job, if you can stop the concrete you've won most of the battle. Centex had gotten a nonunion concrete company to deliver the concrete and the drivers were going through the picket line.

When I arrived the cement trucks were just pulling up. The picketers, who were union members of Local 642, were blocking the truck and beating on it with their picket signs. I joined them at the bottom of a long hill, and by putting our bodies in front of the cement trucks; we turned away quite a few. As the next truck pulled up, a pickup truck came rushing down the hill from the job site. Some guy, I assume the superintendent, charged out of the truck and began screaming at the cement truck driver to run us down.

The cement truck edged forward, gaining speed, and I unfortunately was standing right in front of it. Other picketers began jumping out of the way, but I avoided the tires by climbing up on the bumper, grabbing the bulldog ornament and pulling myself up. I yelled at the driver to stop, but he continued to pick up speed.

To my amazement two carpenters, Mike Casado and Ray Silva, jumped on the steps next to the doors, and yelled to the driver to stop. He wouldn't stop and we proceeded to destroy the front and sides of the truck, kicking any part that might stick out. If the truck didn't stop, it would soon deliver us on to a jobsite full of pissed off scabs.

Just as I thought I was going to have to fight my way out. The Business agent for Carpenters Local 2046, Bruce McDougal, pulled his car in front of the moving truck to block its way. As the truck stopped we all jumped off. The superintendent who had ordered the driver through, pulled up alongside me, opened the door, and reached into the glove compartment for a pistol while I was trying to climb in the truck to kick his ass.

Before he could get a hold of it, the driver of the pickup accelerated back up the hill into the job site. We all jumped into Bruce MacDougal's car and sped away just as the police arrived.

I headed back to San Francisco, knowing I would have to tell Roy Fouche that I had not only failed to calm things down, but had possibly made things worse. What I didn't know was that while all this was going on, two business agents, Bill Linck and Frank Castiglione were at the police station meeting with the chief of police, trying to reassure the chief that everything was under control. They had listened to the whole fiasco over the police radios while sitting in the chief's office. They were not happy campers and were on the phone to Roy Fouche as soon as they found out I was involved.

By the time I got back to San Francisco Roy calmly asked me for the story. I told him the truth, knowing full well I had not done what I had been instructed to do. I took responsibility for what had happened. Roy yelled and screamed at me and sent me to tell Jim Green and the lawyers what I had done. I thought for sure I had screwed up and would be fired - instead I was invited to lunch.

The Homebuilder Strike continued, one of my jobs was to research the employers we were striking. One company was Ponderosa Homes, a fully owned subsidiary of Aetna Life and Casualty. I was assigned by Jim Green to work with Jimmy Larson from the Sacramento District Council of Carpenters to launch a Corporate Campaign against Aetna Life Insurance.

In 1984, the concept of the Corporate Campaign was pretty new. Unions were just getting into attacking the real owners of corporations and targeting them. I knew nothing about corporate structure, stock holders, subsidiaries or holdings.

I did know that the downtown buildings didn't like me creating havoc on their property. Those corporations don't like any kind of negative publicity or visibility. I figured to do the same thing to Aetna.

First I had to do a little research, and the best place to investigate businesses and corporations was the Stevenson library at UC Berkeley. Remember, this was before the Internet and research was done by hand. This was before computer databases.

So, off I went to the Stevenson to research Aetna; to be honest, I had no idea what I was looking for. I had been a horrible college student, clumsy and inept in my work. Somehow I managed to find the "Pension Blue Book," which listed every pension plan and every health and welfare plan in the United States. It also listed the clients of every large Health and Welfare Plan with whom Aetna Life and Casualty did business.

I copied what I could that day; later we bought our own copy of the Pension Blue Book. Jimmy Larson, the Secretary Treasurer of the Sacramento District Council and I began contacting every union who used Aetna for pension or health and welfare coverage. We started by building a mailing list and sending out flyers telling other unions about our fight. I called every union trustee I could find, telling them about the fight, asking them to help, and finding out when their Trust Funds would renegotiate.

Then one day Jim Green came to me and said, he was getting calls from people who were using Aetna and found out they were nonunion shops. It seemed that we weren't the only ones Aetna was screwing.

As the campaign against Aetna Life Insurance grew, we encouraged unions whose contracts for health care and pension were expiring with Aetna to change to another provider. Small unions first responded by moving their money, followed by bigger unions. Our corporate campaign was picking up steam.

Then, came the International Foundation of Employee Benefits meeting in New Orleans. The International Foundation is an organization that serves to educate both union and management trustees on pension and health and welfare matters. Thousands of trustees attend these conferences.

Jim Green had gone there as a trustee to force the Aetna issue on their home turf. He arrived in New Orleans hoping to get an audience with the International Foundations Leadership. He had written and called ahead of time to Lee Pollichek, the President of the foundation. He was served with an injunction by the International Foundation saying the Bay Counties District Council of Carpenters could not handbill or disrupt the meetings in any way.

That afternoon Jim called me from New Orleans and told me to pack my bags and bring all our Aetna stuff. We had been preparing a newsletter for distribution to all Aetna accounts we knew of in addition to tee shirts and buttons.

I bought a ticket for my wife Leslie and the next day we headed for New Orleans with bags of information. The farthest I had ever been from home before was Winnemucca, Nevada.

The first day we were there I began hand-billing the attendees with our literature. I was told by the off duty New Orleans police who served as security guards at the hotel, that I wasn't in California and this kind of "hippie shit" might not work in Louisiana.

The second day we went into the convention hall and began putting our information out on the seats. We were immediately surrounded by security guards and literally pushed out the front doors of the hotel. All the way out I was screaming, "They're throwing a union out because Aetna's building nonunion."

Both my wife Leslie and Jim Green's wife Josephine were startled, scared, and I think a little excited by what took place. I remember Tony Viola, the financial secretary of Local 2046, being thrown against the doors of the hotel; I think he felt it was the most exciting thing that ever happened to him.

The third day, we passed out flyers until about 1:00 p.m., at which time I was ordered into the hotel and convention center. There was Jim Green, John Watts and Victor Van Bourg, our union's attorney. The International Foundation was going to give us time to present our case to the board of directors, which had been Jim Green's intention all along. We had won. Leslie and I went home full of stories and I smelled blood. The attention we received drew more unions into the crusade and began costing Aetna big money.

The work had paid off; the International Foundation fiasco had embarrassed Aetna Life and Casualty in their house. Our efforts in reaching and uncovering the truth about the nonunion practices of Aetna had paid off. Unions all over North America were moving their pension and welfare accounts to other companies. By the end of the campaign Labor Unions around the United States had moved more that 800 million dollars. A few weeks following our trip to New Orleans, Ponderosa Homes signed the Master Agreement.

In a short time I had learned that building owners could not stand the heat caused by disruption of daily activities in their buildings. That the employers were only brave when the union was passive, and third, one of Americas biggest corporations flinched when we started affecting them financially.

I also was working alone most of the time. I found that risk-taking is not a common human trait. Without sounding arrogant, I was doing things others were afraid to take on in fear of embarrassment or failure. Throughout my career I have met many smarter people than I, but few who would put a plan into action. And fewer who could see and plan for a variety of events. I watched and learned as Jim Green would tell me and teach me things and I would watch impatiently as they unfolded.

Michael Munoz – Arrest Railroadway/Hatch

Chapter Twelve: Railroadway/Hatch

In the spring of 1984 the State of California, more specifically Cal Trans, awarded a 47 million-dollar contract for the construction highway interchanges located on Highway 580 and 238 in Hayward California.

Every once in a while some moron for the state will decide that they can save money by awarding work to nonunion contractors. Dan Curtin of the State Council of Carpenters prepared a great study proving union highway contractors build faster with less loss of life and usually under budget.

This project was awarded to RailRoadway/Hatch, a joint venture supported by the Associated Builders and Contractors (ABC). The ABC alleges to represent the non-union construction industry, but my research into their membership shows that half their membership may well be providers of construction services, but do not actually employee construction workers.

RailRoadway was the non –union arm of Riddell International a union contractor. This would be the largest nonunion project ever awarded in Northern California

Unions like the Carpenters, Operating Engineers, Laborers and Cement Masons worked together on job sites, in which the same employers are all signed to the master agreements which were designed similarly by Victor Van Bourg. These unions are usually referred to as Basic Craft Unions. Other unions, such as the Electricians, Plumbers, Asbestos Workers and Sheet Metal were more fractured, each local independent from the others. However, they also tended to have more in common with each other than with the basic craft unions. Because their memberships are much smaller, these unions have learned to be more active in the Building Trades Councils, gaining strength in numbers. They also were prone to long standing feuds, unintentional slights, and raiding of work jurisdiction. To this day these relationships are a mess and perpetrated by the leadership of the State Building Trades Council. It is a wonder that the AFL and the CIO merged.

In many ways the Carpenters unions are a small version of the Building Trades Councils. The Carpenters Union is made up of Carpenters, Pile Drivers, Millwrights, Cabinet Makers, Millmen, Shipwrights and Industrial Workers. But it is always the Carpenters who take the biggest hit from the other trades. With accusation of raiding members, or stealing work jurisdiction. In Northern California the Carpenters have been both the most conservative and the most progressive union dependent on the era. Even though I am a Pilebutt, I refuse to stand by while other trades outside of the carpenters trash them. Time and time again the Carpenters Union in Northern California has stood up to the plate and fought or financed the biggest attacks on the building trades unions in Northern California.

Until now I had pretty much stayed away from the building trades meetings. I had not ever received any help from the leadership of any building trades' representatives. But I have to say it was the rank and file working union man who time and time again honored picket lines we set up on the Hatch project or refused to handle material delivered to job sites.

I was assigned by Jim Green to work the Rail/Roadway Hatch project. My job was not to organize the non-union workers but to affect the job site in a negative way. The Alameda County Building Trades led by Joe Egan wanted to act. At my first Building Trades meetings I found that they where totally unprepared to fight and their leadership was just like the carpenters' agents, some good and some bad.

They hired John Richardson, who formed a company called the Labor Research Group. A lawyer by profession, he had been trained by Patty Gates, who now works for the Van Bourg Law Firm. He was hired by the Building Trades to monitor and enforce prevailing wages on the project. "Prevailing wages" are established by the State of California based on surveys of the average pay for each trade. Employers must pay these wages on all state work.

I set up picket lines based out of Local 1622 in Hayward, California, one block away from the project. We staged acts of civil disobedience and cooperated with the Labor Research Group.

Our picket line was established to inform union members that the job was nonunion. It was our goal to control the quality of the labor entering the job site, forcing RailRoadway to bring skilled workers from outside the area. I already knew that the good construction workers had jobs that allowed them to work near home and peer pressure might keep union men off the job. That meant failure for the employer, union members would not work on the project in the numbers the scab employer needed. RailRoadway was an out-of-state contractor that depended on a few highly skilled travelers to start the project. But they needed local building tradesmen to build the project.

Union members had decided by themselves that they personally would not work on this project. The job had been partially completed by nonunion men with limited skills. Defects in workmanship were uncovered in the bridges and abutments and had to be repaired. I believe in my heart that the fight the Union put up was only half the effort, it was the Union Member who beat RailRoadway/Hatch. This lesson would be proven time and time again as I grew older. At the same time, John had discovered that the out of state contractors on the job had no California State Contractors Licenses, which amassed huge fines.

While we continued to harass the job, we hired our own security guards to make sure no union would not be blamed if any vandalism took place. We were running employers costs up and slowing the project down. At one point we held a march attended by more than ten thousand union members. As support for our cause grew, better relationships began to develop between the Building Trades Unions, local political organizations and politicians. Almost two years to the date the project began, the job went bankrupt while only fifty percent complete. The building trades had found a way to attack the job through the Labor Research Group. The Carpenters had paid their full share plus assigned me full time and paid the picketing expenses.

Remember that during this time, Reagan had beaten the Air Traffic Controllers and unions were reeling backwards. The ABC was banking on a weak economy, an anti-union president and the fractured state of the building trades. For the most part union leaders were old men, caretakers of a dying organizations they sat by and watched shrink. After this fight many building trades union began hiring organizers and becoming more proactive in the fight to keep our standard of living in the Bay Area.

A union contractor finished the largest nonunion highway project ever allowed by the state of California. Although this time, the Associated Builders and Contractors had gambled and lost, they weren't going away.

Chapter Thirteen: *Stewards Program*

During 1984 I was assigned by Jim Green to instruct the Bay Counties Stewards and Leadership Training Program. The program was designed to educate union members about the union. It was mostly guest speakers who talked about current events in the local unions.

Jim Green was not a micro manager and he allowed me the freedom to adapt and change the program. I began holding the classes twice a year, in a different local each time. The classes were a series of 8 three-hour programs based upon the various departments within the union's structure.

Health and Welfare Trust Funds, duties of the local unions, legal matters facing union members, union history, apprenticeship and journeyman training, political and current issues were the topics discussed.

Guest speakers addressed the classes, with expertise in each of these fields. By 1997, I had trained over 500 union carpenters in eight week sessions, one night a week, twice a year. Some locals had eight people attend while others had fifty. If one person showed up the class went forward.

I also got to know almost every active member in the BCDC and certainly every delegate. I never used this to gain any political advantage for myself, which I am quite proud of, but by 2001 it was going to pay off big time.

It was during the Stewards program that I learned about the AFL-CIO Organizing Institute. I became a teaching fellow at these three-days and taught about ten times before the UBC began using the same model for their organizing programs.

By the end of my teaching the Stewards program in 1997, the majority of future organizers and business representatives had gone through these classes, and taken leadership roles in the organization. I take great pride in having instructed these courses and having touched the lives of many of the leaders of the Northern California Carpenters.

Victor Van Bourg – Courtesy of the Van Bourg, Weinberg, Rogers & Rosenfeld

Chapter Fourteen: *Victor Van Bourg*

Victor Van Bourg may be the smartest man I have ever met. He was also one of the most intense and the most complex. For more than twenty years he would be my friend, mentor, and at times someone I hated. Someday, maybe someone will write about his contribution to the labor movement. I'm not qualified, but it's my opinion that without him the Bay Area labor movement would have been in big trouble.

Victor was the son of a Russian Jew who escaped his country during the revolution of 1912. He grew up in the Boyle Heights district of Los Angeles and worked as a painter. Members of his father's painters local union helped put him through college. He eventually graduated from Boalt Law School at UC Berkeley.

In the 1990s, I began videotaping Victor in an attempt to document his "institutional memories" of the carpenters' collective bargaining agreement, the grievance procedure, and the building trades unions in the Bay Area.

Victor's firm represented many of the largest, most powerful unions in the country. He represented almost all the Basic Craft Unions, whose contracts were similar. The members worked together, and in general the leadership was cordial to each other in his presence. Without Victor, no effort would have been jointly made. Everyone trusted him, even when no one trusted each other.

What I really learned from Victor Van Bourg was that it was all right to break the law. The struggle for working people and the decline of the labor movement was due to changes made in the law to keep working people from affiliating with one another.

Lawmakers seldom reward Labor. Republicans seem to always eat away at laws and conditions which make unions grow. Meanwhile, when Democrats are in office, laws they make don't seem to improve or repair conditions eroded away by the enemies of labor. I don't know how many millions of dollars and millions of hours union volunteers waste getting Democrats elected to office without being adequately represented for their enormous support.

I also came to realize that labor attorneys are often very much like the old guard we had fought to change within the Carpenters Union. The Van Bourg law firm controlled the legal interests of the Labor Movement in Northern California. Any labor attorney who was not representing the Van Bourg firm was dealt with viciously, disenfranchised, and driven to other fields of law.

In addition, some labor attorneys tend to take control of an organizing effort, stifling direct action in the field and turning organizing campaigns into armies of picketers sitting in lawn chairs. I recall attending conferences held by the Carpenters International Union where attorneys representing Southern California claimed to have invented an organizing method where union members stand in front of a business and hold up a banner. This is a pretty old and sometimes successful method. But this attorney went so far as to describe how the banner would be built, the material to use, how it was to be held by the banner holder, where they could stand, how many people could stand near it and what, if anything, could be said.

I watched with amazement as organizing directors turned to this one method of organizing, just because this lawyer who had little if any actual experience told them to implement a method that stifled direct action and conflict between owners and the union.

Prior to his death, Victor had been representing the Iron Workers general president against charges of corruption. Facing jail time, the general president testified that he had only done what his attorney, Victor, had instructed him to do. The United States Government initiated action against Victor and the entire law firm. He died returning from Washington, D.C. His name was stricken from the law firm's name as part of the settlement resolution, but no matter what the government ordered, the firm is still known as the Van Bourg Law Firm.

In conclusion, there are very many excellent union attorneys, but only a few good organizing attorneys; the rest need to stay in trust fund meetings.

Chapter Fifteen: *BE&K*

Looking back now, the RailRoadway/Hatch project was the first round by the Associated Builders and Contractors in their attempt to break the building trades unions in Northern California.

Building trades unions have been powerful since the first pioneers walked to California alongside their wagon trains or sailed around the tip of South America. Building trades' workers have always been a dominant force in the San Francisco Bay Area, not just on the job, but also in progressive politics. Rarely can a Republican candidate win in the San Francisco Bay Area.

During the 1980s, conservatives and their Republican allies launched a nationwide attack on labor. They pushed the open shop system and named it the Merit System, based on the American Plan of the 1930s. When Ronald Reagan broke the back of the air traffic controllers, the AFL-CIO leadership under Lane Kirkland did nothing. Big Business and their friends went on the attack, and we watched as factories were bought and sold for their assets. Union jobs vanished in an instant, pension plans were looted, unionized plants were closed, and entire regions of the United States were economically decimated.

All this set the stage for trade agreements such as NAFTA, GATT, the rise to power of the World Trade Organization and the banking collapse of 2008. American workers don't only have to worry that the company isn't doing well, we now have to worry that an undocumented worker will take our jobs or our companies will fail because our Congress signed treaties that protect foreign business interests over American business interests.

Like I said, the Associated Builders and Contractors were the spearhead for the attack on the Bay Area building trades unions. Their plan was put into effect long before any knew or heard of the Interstate Highway 580/238 interchange

March 8th and 9th 1988 – Ten Thousand building trades workers shut down San Francisco in protest of the Associated Builders and Contractors National Conference.

and long before anyone knew about the new steel mill that would replace the antiquated one at USS Steel located in Pittsburg, California.

United States Steel was trying to stay competitive with foreign steel producers. They formed a joint venture with Pohang Steel, the largest steel manufacturer in South Korea. Together they rebuilt the steel mill in Pittsburg, California, and they renamed it USS/POSCO.

They originally went to the Contra Costa building trades and asked for a project agreement. A "project labor agreement" would have been negotiated and signed between USS/POSCO and the Building Trades Unions. The steel mill asked for 90% wages based upon the current wages paid to building trades workers and full benefits, and the job would last two and a half years. These types of agreements are common in big industrial facilities, like refineries, which where all over Contra Costa County. The Basic Crafts were agreeable, but the others, like the Plumbers, Electricians, Sheet Metal workers, Asbestos workers, these unions representing fewer workers but who were higher paid, and who controlled the building trades, voted not to sign the project agreement.

USS Steel had an alternative plan when the building trades voted down the project agreement. The owners had BE&K, one of the largest nonunion contractors, waiting in the wings. BE&K were one of the biggest employers in the Associated Builders and Contractors, and had extensive experience in rebuilding industrial facilities like the steel mill in Pittsburg.

The building trades were immediately fractured and sides were taken. The Basic Craft unions began to meet on their own and develop a plan to take on the steel mill. The building trades once again thought that they could hire someone to do their work, as they had done during the Rail/Roadway Hatch project in 1984.

This wasn't a public works project funded by state taxes. This was a privately funded job, surrounded by fences and as if we needed another problem, there were 600 union steel workers depending on the new steel mill's construction to keep their jobs.

Jim Green called me in one day and told me that I would be assigned to the project. Victor Van Bourg, the union's attorney, was also there. Together they explained to me what was going on. Victor had a plan; he drew it out for me on the placemat in the restaurant we were sitting in. I was scared shitless, because here were two of the most powerful people I knew in the labor movement, and the plan to stop one of the biggest nonunion contractors in America was being drawn out in front of me on a placemat. I was going to be assigned to be the point person for a multi craft effort to stop this project.

I knew something they may not have. I was going to work when others were standing around. I would work night and day, and I knew from the RailRoadway/ Hatch job that if the skilled union men would not work on the job, they would have to import their work force, which would increase labor costs. Pretty simple plan, and with luck, we could take advantage of other weaknesses as they came to light. Not as complex as the POSCO Placemat Plan, but hopefully at least as effective. Keep the union men out and make sure others knew they were staying out.

One afternoon I was called into a meeting of the Basic Craft unions, who were being introduced to the plan but together by Victor and Jim Green--the famous "POSCO Placemat Plan." Jim Green, Victor Van Bourg, Tom Stapleton from the Operating Engineers, Tom Clark, head of the Laborers, Chris Hernandez of the Cement Masons and Dick Zampa of the Iron Workers were present.

Victor started out the meeting describing the plan and how it would be funded; he then described how the effort would be staffed. He explained how the

Carpenters Union had a guy who would be assigned full time. He exaggerated my success and was building to the greatest introduction of my life. Here I was, 35 years old, in the meeting with the leaders of some very large and powerful unions. I was going to spearhead the fight of the decade against one of labor's largest enemies.

Whoa, I thought, stop the cameras. I was so nervous that I had put the pen I was taking notes with onto a paper clip. While my introduction was taking place I began to spin the pen by the clip. The pen was spinning round and round, and I hadn't realized it but the centrifugal force had caused ink to fly out in the end. As Victor's introduction concluded everyone turned and looked at me.

I was covered in hundreds of blue dots and I didn't know it. For a brief moment the looks on their faces were looks of shock, disbelief and awe. I thought they might be impressed, until I saw Jim Green bury his face in his hands. For the first time in his life Victor Van Bourg was speechless. The room erupted in laughter and the fight against BE&K and the ABC was on, with me as the blue dotted meanie at the point.

The project itself was in Pittsburg, California; the plant is the size of a small city and has its own docking facilities. The cement masons had given us office space about a mile from the site. That would be my home for the next two years. The other basic craft unions had assigned business agents in the area to help with the fight. Once again, the Operating Engineers were there, the cement masons supplied picketers as well as the Iron Workers and carpenters.

We manned the picket line rain or shine with twelve picketers every day for two years. Of course many days we would see fifty or one hundred building trades men and women on the line.

The POSCO Placemat Plan was one fought off the site. BE&K had formed a joint venture to build the project with Eichleay Corp. Eichleay was signed to an International Agreement with the Build Trades. But the joint venture was being operated nonunion. Since Eichleay had no men, equipment or supervision on the job they were legally double breasting the job. We had to try and prove that Eichleay was on the job and that the job was covered by the international agreements Eichleay was signed to.

We established our picket line in front of the offices and the main gate to the plant. I had been having success picketing smaller contractors' offices and reasoned that the site of the picket line every day, would nag at the executives who expected the scabs to cross if they wanted to work. To this day I am amazed that the gate for the nonunion contractor BE&K was allowed at the plant's main entrance. There were two other gates, one for the few union contractors who we could not stop bidding work inside the project and the third gate for the steelworkers. They were all at least a mile apart.

Our picketers were all union members and we spent hours training the permanent picketers on the law and procedures we would demand. Very few ever violated our protocols, except in extreme conditions, which always occur once a picket line becomes effective.

As the job began to man up, the nonunion construction workers began to arrive from out of state; our union members did not. If the project was going to get built it would be the first 100% nonunion job built by transient out of state workers.

I also knew that the picket line would stop union members from going to work inside. I also knew that we could find out information about what was going on inside if the picketers asked questions of people leaving. We knew what was going on inside before POSCO did.

The everyday onsite picket line activities were handled by a cement mason name Dave Hansen, who was sharp and quick to learn. One day a truck driver stopped and asked him where Eichleay's yard was. Dave knew we were trying to prove Eichleay had men and equipment on the job, so he asked him to see the bill of lading and took one of the copies as he looked at the paper work. This simple act was one of the major factors that caused BE&K tremendous financial loss. The case was still in the courts being fought and appealed into the next century. Sandra Benson was the attorney assigned to the BE&K campaign and it was her efforts that won this fight.

The picket line also collected information from the nonunion workers as they left quitting or being fired. We contacted injured workers and found them legal representation. In total five workers were killed working on the site and a few others in town on drug-related issues.

Because our picket line was the main gate for the steel mill, this meant the independent truck drivers hauling steel in and out also crossed our picket line hundreds of times a day. In the morning a thousand scab construction workers would cross the picket line, and in the evening the same thing would happen.

The road into the main gate was two-lane with cyclone fence on both sides. Once you passed the main offices, there was a thousand yards of fencing with no way in or out except to go into the plant and turn around. The main gate was a gauntlet; once a car was committed it had to go.

BE&K had built two scaffold towers on company property in which they put security guards and video cameras facing the picket line. One day a picketer was accused of some kind of mischief. BE&K called the police and produced videotape from one of the scaffold towers. The next day the picket brought out a mirror and reflected the sun light into the camera. I'll bet we burned out twenty cameras, and you can bet if something illegal was going to happen then the sun light blinded the cameras at the instant some mischief took place.

From the first day, there was always someone aiming a car or truck at a picketer. Our pickets knew they had to be sharp or be hit. They also knew they were being watched by guards and by video.

The picketers bore the brunt of resentment and insult from both scab construction workers and for some time the independent truck drivers. We had to do something to stop this or someone was going to get hurt. Our plan was to build large demonstrations at quitting time whenever someone screwed with the picket line. There were hundreds of building trades workers employed within miles of the plant. So we planned a barbeque for the next incident, and the first one was a culinary disaster.

When BE&K saw the crowd building outside the gate they held everyone inside for thirty minutes, pissing off the nonunion people who wanted to go home. Then rather than let them exit through the back side of the plant, they forced them to go through the thousand yards of angry union construction workers lining the street. They wanted a confrontation to occur.

The Operating Engineers had demanded that we cook steaks for the attending building tradesman. The demonstration turned into the biggest food fight I had ever seen, and the first riot. Food was thrown and shoved into people's faces, angry words were exchanged. Some idiot decided to get tough after he had gone through the line and stopped his car at the end of the gauntlet. The scabs in their cars were now trapped between the fence and the mob. There was no going forward and no going back.

The Contra Costa Sheriff's Department arrived and began moving the traffic. A couple of scabs were arrested and I learned two things. One was next time we would have hot dogs, and the other was the nonunion guys learned not to screw with the picketers unless they were looking for trouble. I think we had to have one or two more picnics before the scabs began hanging their heads and going quietly through the line every day.

I must also add that following the first food fight, it became easier to man the picket line. The picketers became more militant union members. They were taking part without pay; they were people who worked steady and felt the need to pay back and fight for their future and not rely on what others in the past had left us. Our numbers began to swell from twelve to twenty-five daily.

We won control of the street in front of the steel mill. One day a cement truck inside the plant broke down, and the owner, a nonunion mechanic, sent his sixteen-year-old son through the picket line to pick it up. He cursed the picketers and spit at them calling them names and aimed his pickup at them making them dive to the side of the road. Unfortunately for him the picketers on duty that day were Structural Iron workers from local 378. On his way out they stopped the truck and took his pants down and spanked him like a little baby. God's honor, they spanked him and let him go.

When he got back to the yard, he told his father he had been beaten up. His father got in his truck and headed for the picket line. Two picketers, a father and son, were leaving in their truck as he arrived, so he began chasing them down the old highway in front of the steel mill, hitting the rear of the pick-up as the picketers tried to get away.

At some point the two picketers lost control of their truck and went into a ditch, while the angry father lost control of his truck and hit two elderly women head on. When I arrived at the scene, the women were being taken to the hospital he angry father was arrested and I helped the two ironworkers find the guns they had tossed into the field. To this day I thank God that they didn't shoot the guy, but from that day on, no one fucked with our picketers.

Another problem we resolved was between the picketers and the independent truck drivers who hauled the finished steel from the plant. They would not slow down when approaching the picket line and they would sometimes intentionally swerve, pinning picketers between the fence and the moving vehicle.

Someone made a steel "jack" which resembled the metal toys in the game children play, were you bounce a ball and pick up one or two jacks every time you bounce the ball. Except that this "jack" was six inches long with nine sharpened points. Any way you dropped it the points were pointing up.

The jack was concealed in a bag, which was tied to a rope, which was tied to a pole by another rope. As the swerving truck pinned the picketer against the fence, he would drop the bag between the back two sets of tires. The jack would eat the tire up and then as the rope pulled tight it would pull free. If the police came, they would look in the bag lying on the pavement, and find rotten fruit or a half-eaten sandwich. The picketers were never caught and the truck drivers and the picket line formed an uneasy truce which lasted throughout the remainder of the project.

One day, Dave Hansen, our picket captain brought a guy into the office who was a union boilermaker who had been offered a job as a welder. He offered to help us in any way possible. We asked the boilermakers to give him permission to work inside. I asked him for his phone number and he began sending me detailed written reports about the defective welds and any other problems. He signed his letters, "The Rat Buster". A "union salt" is a person who is sent intentionally to work on a nonunion project to gather information. I don't believe the term "union salt" had even been invented yet, but this guy was incredible and a true union salt.

After being hired he was promoted to the head of quality control for all welding going on. He would attend the weekly staff meetings where the picket line would be discussed and what actions were going to be taken. Incredibly, he had the names, addresses and details of every injured worker or accident that took place. One day he was inside talking to one of the bookkeepers who had just been fired, she was very upset, so he told her where our office was.

She came down and told me what had happened to her and I was very understanding and directed her to the labor commissioner. On her way out I asked her if she had taken any work home with her. She said yes, and if I followed her home I could have it all since it was now useless to her. She gave me every name, address, and Social Security number for every nonunion worker working for BE&K. Something like this could have been the most complete list of skilled nonunion workers compiled, and was worth thousands of dollars.

I spent days typing all the information into my Commodore 64 and we began mailing non-threatening flyers to the homes of every scab on the list, telling them about their rights under California law. I believe it was at that point when BE&K began having trouble manning the project, perhaps because the flyers demonstrated that we knew where their workforce lived.

One afternoon the boilermakers showed up on the picket line with the son of their General President, who is now the Boilermakers General President, and asked our pickets to hold boilermaker picket signs so they could take a picture for their union magazine. I asked them if they could pay the Rat Buster benefits; I showed them all the documentation we had received and I explained how valuable he was. They promised to look into it, but I know he never received a dime or recognition, nor the pension credits promised to him by the Boilermakers General President.

By now conditions on the job site were deteriorating; word was getting out to the nonunion workers that the BE&K job was bad, housing was expensive and they were not welcome. It was at this point that we almost got into trouble.

There was a brash young group of union members who were independent of our effort. They formed their own loose-knit gang that went to the check cashing stores around Pittsburg on payday and started fights with the BE&K workers. One Saturday they met at the railroad tracks just before the steel mill shift change and attacked the cars of the scabs who were working overtime that day.

We had to put a stop to it. We had always taken care of our own business, and been able to defend our position with the police. The Chief of Police in the City of Pittsburg was Lenny Castiglioni, the son of Carpenters Local 152 senior business agent Frank Castiglioni. We would lose the support of the police if we didn't stop them so, Jerry Balmer, the Ironworkers business agent and I began showing up whenever we knew they were going to meet. They were smart enough to understand that what they were doing was dangerous to the union's efforts and they disbanded. However the nonunion BE&K work force continued to have problems both with union members and with the police.

I want to say at this point, Ray Trujillo of the Ironworkers Local 378 is one of the unsung heroes of this fight. He was not their part time, if there was something that had to get done he was always there next to me. One time, I found that the railcars that entered the site did so through a gate that could be crossed by workers, and that there was no guard stationed at that gate.

Ray and I would dress in our suits, put on white hard hats and drive into the steel mill. We would walk around the construction site and steal every piece of paper that wasn't nailed down. The construction site was huge, and we were never asked who we were or what we were doing. I have to admit I was afraid of getting caught, but Ray was unbelievable. He would go into the construction offices of BE&K and take whatever he could get his hands on. We even went into the plant and stole the drawings off the work stations! By the time we left my car was full of work papers and payroll records. Lots of this stuff may have ended up in our Eichleay lawsuit.

I had been instructed by Victor Van Bourg to stay away from the picket line as much as possible, but this was an order I ignored. In one of the lawsuits filed against the unions I was accused of some minor illegal activity on the picket line. The documents stated that they had pictures of me. I was ordered to the court house by Victor to make copies of the court documents. When I saw the picture that was supposed to be me breaking the law, I found it was my buddy Raymond Trujillo, and the charges were dropped.

Our efforts began to pay off in the second year. The job was falling behind, workers were being injured or killed, and the scabs were taking action independent of the union's activities. They were quitting and their replacements were even less skilled than they. At one point, after a nonunion ironworker was killed, ironworkers came forward and out on strike, and we helped them handbill workers going into the plant

While all this was taking place the Contra Costa Building Trades had hired an environmental attorney to fight against the steel mill. Initially they had some successes. But their goal was to close the steel mill. Our goal was to stop the project and finish it under union conditions.

The Basic Crafts unions had early on decided that closing the steel mill would mean the loss of 600 union steelworker jobs. Public support was on the side of the steelworkers, the city needed the revenue, and the steelworkers needed the jobs. In the press the building trades unions appeared to be acting like spoiled children, threatening closure of the mill if they didn't get their way, and we wanted no part of that.

During the last six months of the job, USS/POSCO requested a meeting with Jim Green, who invited me to attend. The head of the steel mill met with us and said that the project was behind schedule. He asked if BE&K were to leave the site would union contractors be allowed to come in and make the steel mill operable? Jim Green said, yes if USS/POSCO would sign a union agreement with the building trades. The mill had one condition, that the building trades would drop all their lawsuits against the steel mill.

A meeting was called and Jim explained what had happened. The building trades said that their lawsuit challenging some environmental permit was about to be decided in their favor. That meant the project would come to a stop and the building trades viewed this as a huge victory. This was a significant disagreement; the Basic Crafts were for the agreement the Building Trades were against. The vote at the building trades was to continue their lawsuit against the steel mill.

The Building Trades lost their challenge in court, and BE&K left town leaving the project incomplete. The picket line came down and union subcontractors went in and finished the steel mill at 100% of union scale. What could have been a resounding victory for Building Trades Unions around the country was won in the end with little fanfare.

I left the fight with new knowledge and with experiences few organizers had.

Chapter Sixteen: *The Bad Loans of 1988*

What began as rumors of bad investments by the United Brotherhood of Carpenters soon became fact. The person who brought them to light was Northern California's 8th District General Executive Board Member M. Bud Bryant.

As Bud Bryant began to question and demand an accounting of the "loan situation" more and more facts came to light. But before Bud could complete his mission he passed away and will be remembered as the first General Executive Board Member who spoke out against the financial mismanagement of members' funds.

The loans had been made under the tenure of General President Pat Campbell, who had been appointed to the position by retiring General President Bill Konya. In a letter to the Local Unions dated September 14, 1989 General President Lucassen, who was appointed by Campbell, announced the foreclosure on loans, which with fees and unpaid interest amounted to almost $95,000,000.

The loans in question were made to developers on properties located at:

Henderson Wharf, Baltimore, Maryland	$29,150,000
Clermont Condominium, Nyack, New York	$17,417,838
Towne Square I, Merrillville, Indiana	$7,625,000
Towne Square II, Merrillville, Indiana	$5,410,848
Butler Care Center, Warrendale, Pennsylvania	$2,837,000
Reddie Point, Jacksonville, Florida	$14,804,196

In a Lucassen letter to the local unions he stated, "On three of the six loans made by former General President Campbell, we saw hundreds of union jobs created and millions of dollars in wages and benefits paid to UBC members in Maryland, Indiana, and New York. On the other three loans, construction never really began."

That letter was signed by:

Sigurd Lucassen	General President
Wayne Pierce	General Treasurer
John Rogers	Secretary
John McMillan	Second General Vice President
Joseph Lia	1st District
George Walsh	2nd District
Tom Hanahan	3rd District
E Jimmy Jones	4th District
Eugene Shoehigh	5th District
Fred Carter	6th District
Paul Johnson	7th District
E Louis Heath	8th District
Ted Ryan	9th District
Patrick Matte	10th District

During the ensuing months letters and accusations began to fly. The General Executive Board split and some of the General Officers began to demand answers and accounting for the loans.

Local Unions and District Councils were outraged and motions from the floors of local unions made by furious members screamed for answers. Union meeting attendance skyrocketed. The most asked question was, "How could this have happened"?

By 1990 the 3rd District Board Member Thomas Hanahan and 5th District Board Member Gene Shoehigh had filed a lawsuit to get an accurate accounting of the UBC losses. Local unions followed up, filing their own lawsuits.

The General Executive Board was in turmoil and a division between the International Union's leadership split the General Executive Board into two factions, those who wanted the bad loans investigated and their names cleared, and those who wanted to downplay the seriousness of the potential loss of $95,000,000.

By early 1991 there were rumors of an opponent to General President Lucassen, and a slate of opposition candidates began to circulate. The rumors became fact, and the slate was finalized led by:

John S. Rodgers	General President*
Jack McMillan	1st General Vice President
Jim Davis	2nd General Vice President
Tom Hanahan	General Secretary*
Wayne Pierce	General Treasurer*
Charles Claytor	1st District
Ed Coryell	2nd District
Mike Beckes	3rd District
E. Jimmy Jones	4th District*
Gene Shoehigh	5th District*
Dick Cox	6th District
Dick Hart	7th District
Lou Heath	8th District*

All these men knew that beating an incumbent General President would be near impossible. The names with an asterisk denote incumbents. They also knew that as incumbent officers they were obligated to protect the assets of the United Brotherhood of Carpenters & Joiners. And last of all they knew that losing meant an end to their careers as union officials.

The Lucassen Team was lead by:

Sigurd Lucassen	General President*
Dean Sooter1st	General Vice President*
Paschal Mc Guinness	2nd General Vice President
Jim Patterson	General Secretary
James Bledsoe	General Treasurer
Sal Pelliccio	1st District
Bill Michalowski	2nd District*
Doug Banes	3rd District
Bill Nipper	4th District
Eugene Judge	5th District
Fred Carter	6th District*
Paul Johnson	7th District
Armondo Vergara	8th District
Ted Ryan	9th District*
Patrick Mattei	10th District*

As you can see by the asterisks, the incumbent officers had split down the middle.

By 1990, the General Executive Board (GEB) was in turmoil, and those who were questioning the bad loans were in revolt. During the following months leading up to the General Convention the GEB never met again.

Most members wondered why anyone of integrity or honesty would have joined the Lucassen team. The answer was simple: an opportunity for them to advance themselves at the expense of those who stood up and fought the bad loans.

Since then, the UBC has foreclosed on all the loans and become the owner of the properties in question. And to this day the UBC still may own a majority of this real estate, worth much less than the $95,000,000 that was paid for it.

The Union Officers who forced the issue of the mismanagement of union funds to the forefront, those men who stepped forward and challenged those who had broken their oath and obligation to protect the membership, were now preparing themselves to challenge an incumbent officer who had all the union's remaining assets at his disposal.

Most regional council leadership teams have forgotten the bad loans. Many of the new breed of leaders were themselves Siggies and are now in leadership positions.

As mentioned at the beginning of this chapter, Bud Bryant died before the General Convention took place. Two weeks before his death he came into my office and gave me three boxes of documents relating to the bad loans. He told me to get them out of the office and keep them safe. Following his death, International representative Paul Richards called on Bud's wife asking about any papers, regarding the bad loans, that she might still have lying around.

Douglas McCarron – Courtesy Doug Puppel United Brotherhood of Carpenters and Joiners

Chapter Seventeen: *Douglas McCarron*

Douglas McCarron was born in Chatsworth, California in 1950. He was the son of a union meat cutter, who taught his unionism to his sons. Doug was married and a father before graduating from high school in 1968 during his senior year and for the next ten years he worked as a drywall hanger joining Carpenters Local 1506 in 1968.

As a union member he attended meetings as an apprentice carpenter and served as an appointed steward through an exemption made by the local union executive board. He walked picket lines, served on committees until his election as Business Agent in June of 1978.

In 1980 he was elected President of Local 1506 and served on the Carpenters Eleven Southern Counties negotiating committee. During 1981 McCarron was elected president of the Eleven Southern California Counties Negotiating Committee and then President of the Los Angeles District Council of Carpenters in 1982.

In 1984 he was appointed a trustee on the Southern California Pension Trust Fund. While serving on the Trust Fund, the union trustees discovered that employers had borrowed monies from the Pension Trust and had gone delinquent in their loans. McCarron filed a lawsuit against the Pension Trust, with the support of the union trustees. His suit was designed to rid the board of Jack Bernard, a management trustee who the union trustees allege to have been behind the delinquent loans and the failure to collect.

The suit was settled out of court, in 1989, by the insurance companies representing the union and management. All the trustees resigned and the Trust Fund was repaid 30 million dollars.

Paul Miller, the Executive Secretary of the Southern District Council of Carpenters and Armando Vergara, an International Union Representative, had recognized McCarron's leadership ability early on in his career. When Miller retired it was Doug McCarron who replaced him.

In interviews with Jim Green, the Secretary of Northern California, it was his belief that when the union found out about the delinquent loans they choose Doug McCarron to be the lead on the lawsuit. While none of the union trustees were guilty of any wrong doing they were replaced in order to clean up the Pension Trust.

In 1987 McCarron was elected Secretary-Treasurer of the Southern California District Council of Carpenters. Following in the footsteps of Northern California McCarron began to restructure the Southern Eleven Counties of California.

Like Northern California the Southern California local unions were quite autonomous, each administered the master agreement differently, and there were no set rules for dispatching out of work members. No standardization of the union hiring halls. And like the northern counties before Jim Green was elected, the district council had little power over the local unions and little support from them. There were no standard wage rates enforced in the separate area, wages were established by each local union and rules enforced arbitrarily by each local union and elected business agent.

Using his power as Secretary-Treasurer and using trusteeship powers , McCarron began to merge local unions under a district council run from the top down with accountable leadership. This was done without the support of the local unions; most resisted and a battled ensued. He appointed business agents, and appointed executive boards to the newly formed unions. He took control of the bank accounts and union assets, transferring them to the district council. Unlike the north where resistance was slight, five of the southern local unions sued to stop the mergers in federal court. With the support of the International Union the consolidation of Southern California continued.

It was a standing joke in California that the International Union believed the Indians were still attacking the wagon trains. While California was changing, so was the International Union. From 1979 to 1988 the Carpenters International Union had four General Presidents. General President William Sidell retired to be replaced by William Konya who served no more than one year. First General Vice President Pat Campbell assumed the presidency and was elected in 1985. Campbell resigned for health reasons in 1988 and left Sigurd Lucassen holding the bag over the bad loans approved under the Campbell administration.

In November of 1991 nonunion and for the most part undocumented Mexican drywall carpenters from the city of El Maguey, Mexico, met with Bill Drurey and Ben Rodriguez, two carpenter drywall agents at Local 2361 in Orange County. According to Marco Hernandez, lead organizer in Northern California, the employers had lowered the piece rates so low that the men could not make a living. In January of 1992 the first meeting was held by the workers and leaders were selected. They were Jesus Gomez, Roy Navarro, Nicolas Munoz, Bernardo Segura, Luciano Salazar and Marco Hernandez.

Recruiting of supports started in Los Angeles, Orange County and San Diego. On May 1 1992 a meeting was called in Antelope Valley, California with 500 nonunion drywallers in attendance. At the suggestion of the union they voted to go on strike June1, 1992.

Striking drywallers would meet at the union halls at 4:30 a.m. every morning, at locals in Los Angeles, San Bernardino, Orange County and San Diego. They would organize into groups and go into the booming residential projects. They would convince the workers to leave the jobs, come to the union hall, have lunch and go back out into the field, bringing more workers off the job.

Four crews of workers became eight. Eight crews became sixteen and so on. The strike grew and grew until the industry was shut down with more than 3,000 men refusing to work.

It is important to note here that the strike date had been set for June 1 for a reason. The Southern California Carpenters Master Agreement was set to expire on July 1, 1992. Tough negotiations were expected but with the residential drywall industry out on strike the union employers wanted nothing to do with a strike in the commercial industry. The Carpenters Master Agreement was settled and voted on quickly, thanks to the efforts of the Mexican Residential Drywall carpenters.

The nonunion drywall employers tried unsuccessfully to bring in Caucasian drywall workers from out of state. But as soon as they went home at night their work was mysteriously undone. Only to be redone the following day.

The strikers and the union began filing wage and hour complaints with the State of California for unpaid overtime and coffee breaks required under California Law at the end of September 1992.

During the course of the settlement talks the employers decided to begin formal negotiations. At the first meeting the Painters Union sent representatives to the meeting, claiming they represented the tapers. Tapers apply the tape that hides the individual pieces of drywall. The strikers who had never seen nor heard from the Painters Union during the five months strike sent them packing.

The negotiations lasted three weeks and a first agreement was reached between the Southern California Carpenters District council and the Pacific Rim Drywall Contractors Association in late November, 1992. As the contract was taken to the men for ratification it appeared that it was going to be voted down because the men were unhappy with the terms of the Health and Welfare proposal. According to Marco Hernandez, "By the time we got back to LA we were told that the Contract had been accepted." As a result the painters union inserted themselves into the fight and tried to divide the workers. The relationship between the Carpenters and Painters union would never be repaired.

This campaign helped solidify Doug McCarron as a player in national union politics and his support sending staff and money were critical in the success of the strike. The strike was won through hard work by union business agents like Jesus Vallin, Bill Drurey, Ben Rodriguez and Mike Olds of the drywall unions and strike leaders Jesus Gomez, Roy Navarro, Nicolas Munoz, Bernardo Segura, Luciano Salazar and Marco Hernandez. The strike was ultimately won by the 3,000 drywallers from little Mexican towns, who had enough and wanted to be members of the Carpenters Union.

Following the 1991 General Convention, Dean Sooter, the First Vice President, retired in 1993, Pascal McGuiness moved up and Doug McCarron was appointed as Second General Vice President of the UBC.

In 1995 Whitey Rodger's charges of election fraud were upheld by the Department of Labor and a settlement was reached calling for a Special Convention in 1995. Lucassen did not believe he could win. First Vice-President Pascal McGuiness left the UBC following charges of racketeering and corruption, leaving that position of General President and the appointment by Lucassen to Doug McCarron.

At the convention of 1995 Sigurd Lucassen retired and nominated Doug McCarron as General President of the UBC. He ran unopposed. For those in the room who had supported Whitey Rodgers in 1991 there was satisfaction in that Lucassen was gone. The issue of the bad loans was still unresolved but the corruption was over and that era of UBC politics came to a close.

In March of 2011 I had the opportunity to ask Doug McCarron why he had supported Sigurd Lucassen over Whitey Rodgers. He stated that after he met with Whitey Rodgers, former General president William Sidel and International Representative Armando Vergara, he felt that Rodgers was under the impression that the job had been promised to him while Lucassen, who had become General President following Pat Campbell's retirement and the announcement of the bad loans, was trying unsuccessfully to resolve the problems resulting from the bad loans made by the UBC .

To McCarron's credit any retribution by the "Siggies", as Sigurd Lucassen's followers were called, was put to rest, but I always laugh when I run into a Whitey supporter who worked for the International that adamantly denies he was ever a Whitey supporter. I knew who they were and would never forget their brave stand.

While many district councils had been merged into larger units, McCarron began consolidating District Councils into Regional Councils. A Regional Council could encompass a state-wide or multi-state organization. New bodies were formed, new leadership was appointed, leaders were retired with pension incentives, assets were combined and elections were postponed until the new leadership appeared ready to run for office.

In some areas the right to elect business agents ended and business agents were hired and the right to vote on union agreements was taken away. By 2007 there were 55 regional councils. The Southwest Regional Council of Carpenters, the largest regional council, was made up of Southern California, Nevada, Arizona, New Mexico and Colorado representing over 50,000 members by 2008, while Northern California, the smallest regional council, geographically represented 40,000 members.

The new regional council structure allowed the membership to elect delegates, who in turn voted to ratify agreements and elect the Regional Council Executive Secretary and Executive Board. The right to elect local union officers still lay in the hands of the membership, but the days of a local union executive board or business agents pushing a district council secretary around were over.

In Northern California the right to elect local union business agents had been changed to hired representation in the 1980s,but as mentioned earlier, the right to ratify the master agreements was taken away in 1997. The long-standing rank and file movement in the north once again began to grow and the "Wildcat Strike of 2000" erupted. President McCarron withdrew his mandate of ratification by delegates and returned to ratification by vote of the members, a rare retreat on his part from a position once taken.

The International Headquarters in Washington, D.C. was sold and the Carpenters Training Center was built in Las Vegas, Nevada. Training apprentices makes up a major portion of each district resources. Northern California had long recognized the need for upgrading the skills of journeyman carpenters. McCarron, more than any of his predecessors, began pouring hundreds of millions of dollars into apprenticeship and journeyman training centers around the country.

Throughout the history of the Building Trades Unions in North America, unions have fought one another over jurisdiction, the right by one union to claim the work as theirs and not the jurisdiction of another union. This fight for jurisdiction is in my mind the major reason wages and benefits for construction workers are not greater than they are, and why there are unions in one part of the country or another.

In the Drywall Strike the Painters Union claimed the jurisdiction of the taping of the drywall. Even though they had made no effort to support the strikers, the painters marched into the negotiations claiming the jurisdiction and right to represent men who did not know them or want them. These types of fights have gone on for years, with the employers encouraging one side or another to do battle using the age-old technique of "divide and conquer". The Building Trades rules protect the smaller weaker unions. Unions fight openly, dividing their memberships from supporting other unions engaged in contract fights. Some unions would interfere with organizing campaigns claiming jurisdiction even though they have no interest in organizing new workers. And worst of all, many unions will not honor the picket lines of another.

During his tenure in office, McCarron has become a lightning rod of controversy, speaking his mind and taking the Carpenters Union in very nontraditional directions.

In March of 2001, the Executive Board of the United Brotherhood of Carpenters voted to withdraw from the AFL-CIO. In a letter to national AFL-CIO President John Sweeney, Carpenters President Douglas McCarron said, "The AFL-CIO continues to operate under the rules and procedures of an era that passed years ago, while the industries that employ our members change from day to day."

The Carpenters International Union continued to reorganize during McCarron's term, shifting resources to organizing, and hoped the AFL-CIO would make similar changes. I have seen nothing to indicate the AFL-CIO is seriously considering changes that would cure these problems, nor do I see any realistic chance that an investment of more time or resources by the UBC will alter those facts.

As a result of the disaffiliation with the AFL-CIO the Carpenters Unions were forced out of the local Building Trades Chapters, but continued to work with them in areas where there was cooperation.

During George Bush's second run for office in 2004, Karl Rove, Deputy Chief of Staff for the president, called Doug McCarron and asked if President Bush could speak at the Carpenters Training Centers in Wisconsin and Detroit. McCarron agreed but clearly stated that the UBC would not endorse the Republican President.

While on Air Force One traveling to Wisconsin, it became clear to McCarron that Bush did not understand the trade union movement. However Bush did ask what was most important to the Brotherhood of Carpenters. McCarron's reply was "Protection of Davis Bacon".

The Davis Bacon Act was passed on March 3, 1931 by Herbert Hoover, due to the Great Depression and a 25% unemployment rate. The intent of the act was to guarantee jobs on federally funded jobs for local workers. Many states have similar "Little Davis Bacon Acts" which require the payment of established recorded wages rates on state funded construction projects.

Bush's response was, "Davis Bacon would remain the law of the land." During his second term in office Republican senators resumed their attack on Davis Bacon. McCarron went back to the President asking if he intended to keep his word. The attacks on Davis Bacon ceased.

Even though the Carpenters Union had pulled out of the AFL-CIO, President John Sweeney contacted Doug McCarron and asked to him talk to President Bush about and upcoming appointment to the top position at the National Labor Relations Board.

The appointment was going to go to a right-wing attorney for the National Right To Work Committee. Working for such a notoriously antiunion organization, his appointment would have had devastating effects on workers' rights and protection. At the meeting the president asked Rove if this were true? Rove stated it was, but the appointment was never made. In both case the dialogue paid off.

By 2005 the Carpenters would become a founding member of the "Change to Win Federation" an organization made up of unions no longer part of the AFL-CIO.

In hindsight, five years have passed since the Carpenters pulled out of the AFL-CIO. Has anything really changed? No. The AFL-CIO continues to drag along, led by mediocre leaders. The "Change to Win Federation" can claim no great success and no "Change".

I remember when the Lathers Unions joined the Carpenters International Union. Deals were made with agents and International Leadership to ease the transition and smooth the way. I know from conversations with Doug McCarron in 2009 he stated, "I would not make deals that benefited those who were in power when their unions failed. If any union wanted to become part of the UBC, they would do so as part of an honest negotiation that benefited the membership."

But one thing is for sure, as a result of the Bad Loan Scandal of 1991, the Carpenters Union has changed. Training of working carpenters has become a greater priority; union resources have been consolidated and redistributed, organizing efforts have doubled and tripled.

Has the membership of the UBC grown? I could say that until 2008 the answer is yes. Since then, no. Have any other Building Trades Unions membership grown? No. Is the Carpenters Union on the verge of extinction? It is far from it, while many of the Building Trades Unions may cease to exist because they will not change. Where will these unions go? I would hope they would go to a union that embraces organizing and rebuild their organization, not to a union that promises their leadership a soft place to land.

I came up in a progressive organization that fought to improve the organization with the cooperation of the majority of the membership, and an active dissident movement, a difficult task. In the course of my tenure I fought two General Presidents, at the convention in 1991 and along with my local in 1997.

Douglas McCarron came up in a conservative organization and may have used more heavy-handed techniques, but we achieved a common goal, leaving the United Brotherhood of Carpenters intact and sound, so that those who follow will have the resources to continue the fight.

The consolidation of the Northern California Carpenters Unions into the Northern California Carpenters Regional Council in 1997 was expected and supported. But the choice by Doug McCarron of John Casey, the contract administrator for the Northern 46 Counties Conference Board, made little rational sense. As Green was the chairman of negotiations and had been urged by both Gary Martin and Leon Wagner to stay on until the new Master Agreement was negotiated.

In 2011, when I finally had the chance to ask the question "Why did he appoint John Casey instead of Jim Green?" He stated that it was his understanding that Jim Green wanted to retire and that his health had been suffering and that John Casey was his choice to replace him. He also felt that Casey had been a secretary treasurer of a district council, which gave him the needed experience.

When Douglas McCarron took office in 1995, I was excited and hopeful that the old regime was gone. By 1997, I would be in open conflict with the International and fighting alongside my friends against the International takeover. In 2001, I would be back having won what could be won with Local 34 intact and Bob Alvarado elected as Executive Secretary Treasurer to the Northern California Carpenters Regional Council.

The question that needs to be answered is, how we got to this point in our history, fighting nonunion employers, and a government and political system dependent on our money and manpower but unwilling to legislate improvements for American workers. The unwillingness to resolve problems among local unions, international unions, the AFL-CIO and Change to Win, leaves the Carpenters Union isolated like all the rest of American labor unions to "just survive".

At first, the question I thought I wanted to answer was, "What kind of job has Doug McCarron done over the past 16 years?" In reality that question is irrelevant, what matters is will the Carpenters Union survive and the answer is yes; I am not so sure about many of the others!

Chapter Eighteen: *The UBC Picks Sides*

By early 1990s the loan scandal was rocking the UBC. The leadership of the UBC was dividing up into various factions. The two main factions were the Siggies and the Whiteys.

The Siggies, or those who supported the general president, were made up of internal UBC staffers who lived in the national headquarters. Many were International representatives. It was very common for International representatives who served a particular General President to one day be moved up into the District General Executive Board seat, not for any particular reason or achievement except as recognition of their service. There were District Council Secretaries; almost all of whom were from the small, less powerful districts, who generally had an axe to grind against the larger district councils

Then there were the Industrial Councils, a segment of the UBC who were pretty much left alone, except that their membership base was continuing to erode as a result of neglect, or the industries leaving the U.S. As it turned out, however, they were an unknown force in the International.

These councils had historically been a thorn in the side of the International Union. In the early part of the 1900s, Lumber and Sawmill Workers left the UBC by the thousands to organize for the CIO. Those who remained were stuck with company unions. Don Cameron, a member of Pile Drivers Local 34, was an International Representative who followed the CIO organizers throughout Northern California negotiating sweetheart contracts where ever the company was willing to accept the Carpenters Union as the lesser of two evils. The CIO unions were believed to be controlled by communist factions and the Carpenters Union was a known Republican Union. These bargaining units generally were looked upon as a local union.

By 1991, Lucassen would combine hundreds of small local unions into enough votes to become a decisive ingredient in the election. I know for a fact that Local Union 2882 in Santa Rosa, California sent delegates to the 1991 General Convention even though the local had decertified from the union and the charter represented zero workers. Somehow these delegates were seated at the convention even though no elections had taken place because there were no officers or members in that local.

The Whiteys were made up of the District Council Secretaries from the large district councils. The strength of the UBC still presided in the large metropolitan areas, such as the San Francisco Bay Area, New York City, Chicago, Detroit, Cleveland, Seattle, Kansas City, and Louisville.

These District Council Secretaries represented almost 70% of the UBC membership. They were experienced in large-scale negotiations, they understood large local unions. They represented carpenter contracts in the largest cities and economies in the world. But through political device these powerful district councils did not represent a majority of the votes of attendees to the General Convention. The major players in this drama spanned the continent. As mentioned above, the teams were drawn and the players took their positions.

Jim Green was the Executive Secretary of the Bay Counties District Council of Carpenters, and his story is at the heart of Chapter 8. At the time of his forced retirement in 1997, by Doug McCarron, Jim Green had been elected five times as District Council Secretary.

Both sides, Lucassen and Whitey Rogers, approached Jim Green with offers of the 2nd General Vice President seat. Jim declined, unlike other leaders he went to the floor of the Bay Counties District Council and the membership voted to join the Roger's Team.

An interesting note at this point is that while Northern California backed Rogers, Southern California was backing Lucassen. Both Jim Green and Doug McCarron, the Executive Secretary of the Los Angeles District Council, had agreed that this fight at the International was not going to divide California.

Only one District Council Secretary of the large councils came out in support of Sigurd Lucassen, and that was Douglas McCarron of the Los Angeles District Council of Carpenters. Doug McCarron, discussed in Chapter 17, had come out of the drywall industry and quickly moved through the chairs, from business agent to executive secretary.

Jim Green had always supported and promoted Doug McCarron, a young and aggressive leader from Southern California. It was known throughout the UBC that Jim Green was Doug McCarron's friend and mentor.

Whitey had the support of the large district councils, Lucassen the support of the small district councils. The election was too close to call.

At the 1991 convention 400 elected delegates from the Industrial Unions arrived in Atlantic City. They were lead primarily by Mike Bledsoe of the Lumber and sawmill workers. The Industrial Unions were all broke. Lumber and saw mill plants were closing or decertifying. Reaganomics and corporate raiders had decimated the industry. During the "Louisiana Pacific Struggle", Northern California carpenter locals donated millions of dollars in support towards striking LP workers. Bledsoe's reward for his support was the job as Secretary Treasurer of the International Union.

Mike Draper, Bledsoe's protégé, was appointed the Executive Board Member of the 7th District upon the retirement of Paul Johnson, a long time International Representative.

By the time they arrived in Atlantic City for the 1991 Convention they had with them every delegate they could dig up, paid for by the United Brotherhood of Carpenters and Sig Lucassen. They were all boarded up in one hotel and watched night and day until Election Day when they were marched in lock step by Mike Draper down the Atlantic City Board walk to cast their votes for the Sig Lucassen Slate.

Draper's assignment was to baby-sit the Industrial Union delegates at the 91 Convention. His responsibility was to make sure no "Whitey" supporters got near the Industrial Delegates. Those Industrial Delegates were housed in one hotel and were lead in grand style to the Convention Center on Election Day, lead by Sig Lucassen, James Bledsoe and Mike Draper.

Mike Draper is currently the International Vice-President of the Western Region. The region serves more than 30% of the entire UBC membership. I have to say that he has done the job asked of him and has put the past behind him. He has been tough, competent and fair.

As court records latter proved, the Lucassen slate had used UBC money to pay the way for the broke industrial unions to send delegates to the General Convention, a violation of federal law but enough votes to win. Without the Industrial Union delegates, Whitey Rogers may have emerged as the winner in the 1991 Convention.

Chapter Nineteen: *The 1991 General Convention*

In the spring of 1991 I was asked by Whitey Rogers to manage his campaign efforts. Politically this would determine my career. If I had aspirations of moving higher in the organization, the outcome of the election would be decisive.

Fortunately for me, I had made the decision a long time before this election to remain a union organizer. I had watched as others used organizing as a stepping stone to advance their careers, moving up through the chairs into leadership roles.

I've also seen time and time again in the elections for building trades Secretaries, incompetent men elected into positions they were not qualified for, to make room for someone else lower down the ladder. I took the job knowing I could become cannon fodder, but I also trusted Jim Green to have my back if things went south.

I decided to run the campaign like a bottom-up organizing effort. My first duty was to build the committee. I knew from experience that the Council Secretaries would not be able to put the time required into the effort. So, we appointed "Whips". They would grade each delegate according to his or her support.

We identified district councils where the executive secretary was a supporter; then we identified supporters within those districts. Where we didn't have support we found supporters. Then we identified each delegate and if they were a supporter or not.

By the time we got to Atlantic City I thought it was a dead heat, I had no idea that the industrial unions were attending, and that our that list was short by four hundred delegates. On the first day the Whitey slate marched off the podium in protest. On the second day we shut the convention down for four hours, stopping to relinquish the microphone to Whitey Rogers.

By the end of the first day I had seen more guns than I had ever seen in my life. There were Atlantic City Police, hired off duty police from other departments, hired security guards, personal body guards, and rank and file carpenters who thought a gun would be needed.

We had established our control center in the back of the convention center in the bleachers provided for guests. The International Union would send spies up there to listen and take notes. I was amazed at how many old friends wanted to come and hang out up there, and how many times they whispered under their breath that if "Sig" won the election, I would be fired.

Using walkie talkies, I would communicate with each whip. By the end of the first day someone had received a threat that I was not safe. The Philadelphia pile drivers from Local 454 took it upon themselves to watch my back wherever I went from then on. I will always remember them for this; the convention was serious business and it appeared that the Siggies were getting mad. I later was told that the New York District Council had assigned a "private contractor" to watch my back. I don't know what that means, or if it's even true, but I am glad for whoever was there.

On the second day we began pounding on tables and banging on the tops of straw hats. For hours no one could speak. Following a fist fight on the convention floor, Doug McCarron complained to Jim Green about the noise and the possibility of a riot. Jim looked up and signaled to me from the floor to stop the noise. I got on the walkie talkie and within seconds the convention center was quiet. The look on Doug McCarron's face was almost worth the ass kicking we were about to get. At another point Whitey Rogers ran out of time at the podium. The chairman immediately cut off his microphone. Whenever the Chairman wanted a Siggie to speak from the floor he would call on a microphone where the speaker was wearing a pink baseball cap. We had anticipated this move and I will elaborate further on in the final chapter.

On the morning of the third day we used crickets as noisemakers, those little metal clickers we all had as children. On the afternoon of the third day George Vest of Chicago read a written statement, and Dean Sooter the Vice President responded as though he had been prepared or had the statement hours before it was read. Now momentum of the convention moved firmly into the Lucassen camp.

To this day I believe Victor Van Bourg had given them the statement prior to George Vest taking the floor. As Sooter responded, I got on the walkie talkie and asked for Victor by name. He responded by saying not to use his name on the airwaves. I would not be surprised if he played both sides against one another. His goal was to become General Counsel for the Brotherhood should Whitey Rodgers win. As shrewd as he was he could see the tide turning and was beginning to distance himself from the losing side. He was at every strategy meeting I attended, and when I accused him of it he jumped down my throat.

Nominations for officers were scheduled for Wednesday afternoon, and the Philadelphia District Council had bused in a couple hundred members to watch. The Lucassen Team, however, refused to let them into the convention center. Police with dogs were called out and the members were getting quite ugly standing outside.

All our delegates had been instructed to request guest badges even if they had no guests attending the convention. I took the guest badges and began handing them out to the whips and the Philadelphia members. As the balcony area began to fill, panic set in to the Lucassen team. We handed out about three hundred badges, as I was handing out the last of them one of the Siggies caught me and had an Atlantic City police officer stop me. I passed off the last of them to another whip one of the New Jersey whips. Jim Green stepped in between the cop and me and I escaped.

Thursday was Election Day. As we arrived at the Convention Center, Bob Klien from the Bay Counties District Council, Ron Pettingill from the New York District Council and I saw the entire Sigurd Lucassen delegation marching down the Atlantic City Boardwalk towards the Convention Center. The industrial carpenters would not have even attended the convention under normal conditions. They were broke and being subsidized by the International Union. Their delegates were told that if they were caught leaving the hotel, they would be sent home. They were also told who to vote for.

Pettingel waded right into the middle of the pack asking them to support the Rogers Team. Klien and I followed suit. We were swimming against the tide as we tried to persuade them.

As voters entered the polling place they were escorted to the election booth. The Lucassen Team was a slate, so the voting machines were set up for voters to pull one handle to vote the slate. The union's election committee, all Siggies, could hear if the slate lever was pulled or if individual levers were flipped. Voters were told that the election committee would be listening. Every full time representative was aware that the Lucassen machine was holding him or her accountable. There were also some unresolved questions regarding the ownership of the voting machines. They allegedly were owned by one of the Lucassen team's relatives.

Sigurd Lucassen and his team won the election by a very close margin and the Whitey Team retired or found a place to land and wait.

In the end of this historical footnote, Doug McCarron emerged as a player in National UBC Politics. And to this day no accounting for the bad loans has taken place.

Chapter Twenty: *Director of Organizing*

It was during 1994 that Jim Green began to make changes in the Bay Counties District Council of Carpenters. He had for some time been planning to retire. During a ride home from Sacramento late one night, he told me that he had decided that Bob Alvarado would be the person to replace him.

He asked me if I would take the position of Director of Organizing. It was a position I had hoped to hold some day, so I accepted. Over the past few years I had been given special projects. The modular furniture industry, Aetna Life and Casualty, RailroadWay/ Hatch, USS/ POSCO, and two other organizing efforts against contractors Stokes Russell Hayden, and the Super Structures. In addition, I played a role in the fight over the bad loans at the convention of 1991. I had also been running the Bay Counties District Council of Carpenters Stewards program. For many years I was frustrated by the slow pace in which we were accomplishing our goals.

As part of the special programs I was directing, I was allowed to select organizers to work with me. Ron Spikes, Donna Levitt, Bill Feyling, and others would become the future leaders of the Bay Counties District Council. Our successes would become the models for the United Brotherhood of Carpenters and Joiners of America.

As director of organizing I restructured the department. Rather than have organizers assigned to local unions and doing job reports on union jobs, I believed that it was necessary to target individual nonunion employers; that our goal was to isolate the owner and create havoc on a day-to-day basis which would require the owner to concentrate on the union activity rather than on the his day–to-day work.

The reason most nonunion employers succeed is that they pay less than union scale. Material costs are the same, so the greatest savings a nonunion employer can achieve is by paying his workers less than his competition. Common sense tells me that if an organizer can prolong the job, the contractor spends more money on employment costs, equipment rental and deadlines required by his contract. If the organizer can cause the nonunion employer to hire an attorney and run up costs it will become unproductive for him to continue bidding work.

If the union can slow down his job and distract him from bidding future work and make him deal with the union activity instead of focusing on the completions of these contracts, the union will win. The goal of any union organizer is to become the main focus of the nonunion employer's daily activity. I believe that the organizing department is an extension of the political arm of the organization. Every effort should be aimed at the campaign's ultimate end, having the nonunion employer sign a union agreement or putting that targeted employer out of business. Furthermore, the Regional Council Secretary must be the final authority over all organizing department campaigns. In the same manner that the organizer has authority over his pickets and day–to-day operation, the Regional Council Secretary is the ultimate negotiator and peacemaker. The organizer serves the union, not vice versa; the union always has priority over the organizing campaign. The organizer is a specialist, and his horizons may not stretch beyond what appears as necessary for carrying out his organizing tasks.

Additionally, the organizer must be able to focus and listen to reason in the midst of intense excitement, not allowing passion to overcome his ability to reason and implement his plan. An organizer also must have the self control to balance his intellect, passion and personal ego to achieve his goal without destroying himself, his campaign or become politically injured.

The organizing department must be manned by dedicated unionists who believe in the principles of trade unionism. They must work at the trade, they must be indoctrinated to follow the leadership of the organization and accept its principles.

The organizers must be perceived to be equal in status and pay. The union organizer should be dealt with in the same manner as business agents. This system creates consistent treatment to all employees regardless of job classification and the title. The organizing department should be run by a director who can make decisions based on the information provided by the organizer so as not to cause said organizer to hesitate. In 1978 the district council delegates voted to change the culture within the Bay Counties District Council of Carpenters. I hoped that someday an organizer trained by the Bay Counties District Council Carpenters would someday serve as Executive Secretary

Chapter Twenty-One: *Consolidation of Northern California*

Late in 1996, General President McCarron told Jim Green that he intended to merge the Sacramento District Council of Carpenters with the Bay Counties District Council of Carpenters.

At the Health and Safety Conference held in Palm Springs during the spring of 1997, General President McCarron called the three District Council Secretaries from Northern California together and told them there would be one Regional Council in Northern California. Gary Martin, the Secretary Treasurer of the Central California District Council had convinced the general president that he should complete the merger in Northern California and form a regional council. Martin knew that if the Bay Counties and the Sacramento District Councils merged he would be left alone and isolated and forced to merge under conditions he could not control.

The idea of a merger was no surprise, as the general president had made his intentions known and the idea was supported by the three district councils. What was surprising was the Gary Martin was so willing to take part in the consolidation of the three councils. Bob Alvarado, who was the administrative assistant to the Secretary Treasurer and had been picked to succeed Jim Green upon his retirement, was chosen to develop an Internal Organizational Plan. Staff member Paul Cohen was to work on a representational delegate body and the amending of the bylaws to conform to the mandated international union bylaws and I was instructed to develop the Organizing Plan.

This was nothing new because we had done the same planning when the North Coast District Council of merged into the Bay Counties District council in 1988.

Following the General President's instructions, the three Council Secretaries met. Bob Alvarado and I were called to one of these meetings and questioned by Gary Martin of the Central District Council and Leon Wagner of the Sacramento District Council.

Both Martin and Wagner stated that they were not interested in the new Executive Secretary position. They asked me if I would work for John Casey, the Director of the 46 Counties Conference Board. I stated that I would probably work for whoever was elected by the delegates, but the job should go to one of the elected district council secretaries, Jim Green, Gary Martin or Leon Wagner. I further stated that Jim Green was the person I would support.

In my presence they asked Bob Alvarado to decline to run for the Executive Secretary position if Jim Green chose not to run and let John Casey run unopposed. Bob said no, that Jim Green was the person we elected to run our district and he was the person Bob would support.

I stated that evening that the delegates to the new Regional Council needed to vote on who the new Regional Council Secretary would be. Martin's reply was that the decision would be made the three council secretaries.

I was amazed that John Casey had any interest in the position since he had, a year before, nominated Jim Green for District Council Secretary, praising him for the great job he had done for Northern California over the past 15 years. He praised his negotiating skills, his ability to build consensus, and how Jim Green was recognized across North America as one of the UBC's brightest and most respected leaders.

I knew for a fact that Jim Green had asked John Casey to come to the district council as his Administrative Officer and upon Jim Green's retirement Jim would support John as Secretary Treasurer. John Casey declined, stating that if he wanted the job he would run for it. Sometime later the job was offered to Bob Alvarado and he accepted.

John Casey was a capable man. He was for a short time a business agent in Napa, California and had run for and won the position of District Council Secretary of the old North Coast District Council which he held for a short time.

A few years later he was offered the position as assistant to the Director of the 46 Counties Conference Board. When the Director Larry Bee retired John took his place.

So here is a man who had declined the job, failed to run for the position, and had nominated Jim Green the year before.

The Bay Counties plan was comprehensive in that it called for a large delegate body which outnumbered full time staff and was based on the number of members each local union had. The Sacramento and Central Districts were working together; they developed a plan that was four sentences, basically said to merge all the district councils and merge all the money into one district council. Make the delegate number smaller, giving half to the Bay Counties and the other half to the two remaining district councils.

Gary Martin had been around as a full time officer for many years and was as politically shrewd as anyone I had ever met. His path to the top was littered with the bodies of District Council Secretaries who had confided in him, promoted him and were then betrayed by him. Bill Krebs, the District Council Secretary would lose to Martin after hand-picking him. When the Santa Clara District Council merged with the Monterey District Council, Martin undermined Skip Landry until he had his job. During the 1991 convention, Martin told me he supported Whitey Rogers and then got off the plane in Atlantic City as a Sig Lucassen supporter.

Leon Waggoner was the District Council Secretary of the Sacramento District Council. He had won the seat from Jim Larsen. Waggoner had been a business agent from Redding, California and had never run a local larger than 500 members. Under his leadership his local union grew smaller every year.

We attended the Legislative conference in Washington, D.C. in 1997 and I was called into a meeting with General President McCarron and Mike Draper, the Eighth District Board member. Doug McCarron asked me what I thought of the consolidation and who should be appointed until an election could be held. Once again I said that I thought that the job should go to one of the Secretary Treasurers who had been elected to the position. I stated that the membership in the Bay Counties was still uncomfortable with appointing business agents as opposed to electing them. It was my opinion that unless one of the three elected council secretaries was appointed it would be hard for him to gain enough support to keep the job.

I further stated that both McCarron and Draper knew that Jim Green was the most qualified person. Jim Green sat on the executive board of the California State AFL-CIO. He was chairman of the negotiating committee, and both Martin and Waggoner had begged him not to retire before the new master agreement was to be negotiated. Jim was also head of the Carpenters Trust Funds, the State Council of Carpenters and many other positions of leadership in Northern California.

I told them that I worked for and supported Jim Green and did not know if I would stay working for the UBC if he was not appointed interim to the job. It was because of Jim Green's support and vision that our Organizing Department was one of the best in the UBC, years ahead of the other two district councils and ninety-nine percent of the rest of the country. Efforts to organize employers we were after often failed because the other district councils would refuse to join a fight or would undermine our organizing efforts against nonunion employers.

It was also during this week that I began to notice the stress. For almost three months I had worked daily on the consolidation and directed the organizing department. I believe my intentions were always moral and dedicated towards a stronger union. Remember, while the carpenters had voted to hire representatives, Pile Drivers Local 34 still elected their representation.

It was evident to me the General President was playing games with the lives of the membership of Northern California. He had the power to act and was hesitating. My entire reasoning power and logic spelled out that Jim Green was the right person. But McCarron was acting like these two relative lightweights Martin and Wagner were somehow viable players.

Jim Green had helped make Doug McCarron, he had supported him in every way as he developed from a business agent to General President. They had taken opposite sides in 1991, so that no matter who won California would be represented. Doug McCarron knew every strength and weakness of both Casey and Green. I just knew that logic and common sense would win the day...but not this time.

When we returned from the legislative conference I found that the Sacramento and Central District Councils had launched a "Beat the Bay" campaign. In hindsight, I now see that the deal was cut, and long time friends of mine became distant. Many would not even stand near me when Casey was in sight. Casey was dividing the pie up, trading important positions for support. The Bay Counties District Council was now the enemy. I was sick about it; people were trading the stability of the union for position and prestige.

In May, 1997 Jim Green was called down to the restaurant at the Hilton Hotel in Oakland there Doug McCarron and Board member Mike Draper informed Jim Green that they were consolidating Northern California into a regional council. John Casey would be the Executive Secretary.

Bob Alvarado and I were called down to the hotel and we met with McCarron and Draper. McCarron informed us of his decision. He then dropped a bombshell on me. I was appointed President of the Interim Regional Council and Bob was appointed to the executive board. I felt like I had been hit with a brick and then run over by a car. I accepted because I didn't know what else to do. When I returned to the district council offices Jim Green was in his office talking to Victor Van Bourg. Victor was asking Jim if he wanted to fight the International. Jim stated that he would not fight because he feared retaliation on the Bay Counties members. I walked across the hall to John Casey's office and Michael Rogers, one of the attorneys for the Van Bourg firm, was congratulating John Casey.

I had just watched Victor play Jim Green for a fool. Michael Rogers would not have congratulated John Casey without Victor's blessing. I always felt that Victor and Michael Rodgers were working both sides of the fence. By telling both parties they were supporting them, but in reality working to preserve the Carpenters Union as their client. I was angry at Victor, and I despised Michael Rogers and even more John Casey.

On the second day of my presidency I chaired the meeting of all full-time representatives, but all I remember was a picture being taken. The third day, I chaired the first executive board meeting of the new Regional Council.

The new interim bylaws were voted on and then a motion was made by the Casey Team to buy out a group of business agents. This would mean giving them up to three years of International union pension credits if they would leave quietly. I went temporally blind with rage. Give free pension credits to people who already had enough time to retire, the same people who sold out Jim Green for a few coins of gold/pension credit? What would the members of Local 34 think of me and what about the hundreds of thousands of dollars it would cost?

As president I had no vote, but I did have the gavel and I used it. I spoke against the motion and Thom Donnelly of Local 34 and Mike Knab from 217 voted against the motion, but it passed by a majority.

By Friday, I was without sleep for the third day straight--no exaggeration--no sleep for four days. I went home and talked to my wife Leslie. I didn't know how I could go against everything I believed to be right. The biggest district council led by the best person, Jim Green, the man who had given me so much, the man I respected most was cast aside by Doug McCarron after he had supported Doug McCarron throughout his political career.

On Friday, I walked into the office and passed out, hitting my head on my desk as I fell. By Monday it was clear to both Leslie and I that I would not stay. I could not go against my moral conviction, turn on my friend or explain to the membership at local 34 how I had chaired the meeting where business agents were given free pension credits, while some of them could die before they got the five years needed to vest their pension.

I also knew that as President of the NCCRC I could not stand by idly while the ability of Local 34 members and the carpenters to ratify their master agreement and elect their business agents was about to be taken away from them. I could not live with myself if I was perceived, by those I cared for most, as the person who might take those rights away.

I wrote my letter of resignation and submitted it to Michael Roger and left. By the time I got home Victor Van Bourg was on the phone screaming at me telling me how I had let down the membership. He played on my ego but I knew I had done the right thing.

Chapter Twenty-Two: *Starting Over*

I had entered the twilight zone, nothing was clear to me mentally. I was in opposite land, where everything that made clear sense did not apply. Many of my coworkers called and begged me to return and others were ecstatic about my departure.

I had no idea what I would do, but Leslie reassured me that we were fine. If it weren't for her love and confidence in me I think I would have given up, on myself and on the union. No injury has affected more that quitting the United Brotherhood of Carpenters; I was giving up something I loved very much.

For a few days following my leaving, I received phone calls from Mike Draper and Doug McCarron with job offers. I was amazed at their insensitivity; it was clear what I thought about their actions. One day Leslie made it very clear to Mike Draper, when she cut him short and told him what she thought of him.

To my surprise and amazement, I received a phone call three days after quitting the Carpenters Union. It was Bob Lawson, the West Coast director for the AFL-CIO's Organizing Institute. I had started using the AFL-CIO program when I developed the hiring policy for the Bay Counties District Council.

Bob said he had heard that I had quit the Carpenters Union. He asked me if I would be interested in working for the AFL-CIO. I said yes. He told me the AFL-CIO was looking for a State Director for California. I went through a series of interviews with Mark Splane, the director of the Western United States.

I have to admit that mentally I was a mess and Mark Splane's description of my responsibilities was never clear to me. Then there was a series of interviews and a trip to Washington, D.C. for my interview with the head of the department. That meeting was followed by a face to face meeting with John Sweeney, the head of the AFL-CIO. My responsibilities were still unclear.

A month after the interview I was called by Splane and told that I had the job as State Director for the AFL-CIO in California. There was one small problem; Doug McCarron would have to OK me being hired. I figured that wouldn't happen in a million years so I began making plans for my future outside the labor movement.

I had thought of returning to pile driving, but the regional council had moved to take over the pile drivers union.

After some time I called Monte Byers the communication director for the Brotherhood of Carpenters. We had been friends for many years and he interceded and finally I was given Doug McCarron's blessing to work for the AFL-CIO.

By the time I started working for the AFL-CIO I had received quite a few offers to work for other unions. One in particular was from Machinist District 190 in Oakland California. I was also scared shitless because Mark Splane had totally confused me about what I was going to be paid to do.

I had always looked down upon those people who went to work for the Building Trades, Labor Councils and AFL-CIO. Most were nice people but few had personality to make change. Meanwhile, to this day, the labor movement continues to decline and the labor bodies continue to promote into positions people who have been fired or quit, people who had never led big unions, or people who had grown old and obsolete and were moved into these jobs to make way for younger leaders down the ladder. Now I was one of them. I couldn't take the prospects of being ineffective, or having to deal in a political world of ass kissing and butt sucking that I perceived was going to take place.

Now as I write this book I realize how right I have been. Casey was a complete disaster, but moreover the AFL-CIO has become a laughing stock, the Building Trades leaders are up to their neck in the Ullico scandal, and the national leadership of the American Labor Movement has forgotten their roots. While I may disagree with Doug McCarron's intentions he is absolutely right to refuse to belong to the AFL-CIO. It is broken and needs to be fixed; only radical change and new leadership will stop it from sinking. I believe that the American Labor Movement only survives because the working men and women, the union members, refuse to stop paying their dues.

After three days of employment as State Director of the AFL-CIO I went to my friend Don Crosatto who had offered me work as director of organizing for Machinist District 190 and applied for the job. On Jan 1, 1998 I started my new life as a Machinist.

Chapter Twenty-Three: *Machinists District 190*

The front of the building was entirely glass and every machinist in the plant could see me. So, on days when I would handbill the employees as they left work, I would stand alone on the brick wall outside the plant and wave a five foot by five foot UNION YES banner that was attached to a six foot long pole in one hand and a bullhorn in the other.

The campaign was heating up and the company was trying to win support of the employees. They had been holding captive audience meetings and were planning a big Bar-B-Que for the following day. They had announced it as a day of appreciation for all their hard work.

I wanted to make them focus on the real reason the union had been asked in to organize the plant. Five minutes before the day shift was to get off, all the supervisors including the manager Don Sweetnam, came outside and began hollering insults at me, in an attempt to humiliate me in front of the men.

One of the supervisors, a guy named Dave Henderson, who weighed about 350 pounds and was one of the most hated by the men, began taking pictures of me as the men left the plant. A crowd of about one hundred workers gathered as the supervisors screamed insults at me.

I had to stand my ground, challenging them about the poor conditions and benefits in the plant, them yelling and me using the bullhorn. Then one of the them yelled out, "Go to hell you fat slob!" and I responded, "I'm not fat, I'm portly, Dave Henderson is fat." The workers yelled and cheered, laughing at poor Dave Henderson. Even the supervisors began to laugh at him until the Don Sweetnam began yelling at them to stop.

I went to work for the International Association of Machinists District 190 in January of 1998. Don Crosatto was the Director of Organizing at the time. He had stayed in touch with me since he had heard that I had quit the Carpenters Union. Mike Day was the head of District 190 and was someone I respected a great deal.

At my interview I remember that Mike Day and Joe Bobo, the head of local 1546, were both present. Not only was Mike Day a good man, Joe Bobo was also someone I had tremendous respect for.

Local 1546 was the same local union my father was member of. In 1983 my father had been laid off from American President Lines. He was 62 years old with little prospects for finding work. Many people were on layoff because of the economy. He had decided to retire and collect his union pension; his health insurance had run out and my mother could no longer work because of her inherited diabetes. Their doctor bills were beginning to grow even then. I knew that if he retired without health insurance they would probably lose everything they had worked for.

I went to Local 1546 and asked for a business agent. I met Joe Bobo for the first time and explained to him my father's problem. If he could work long enough to get his health benefits current he could retire with health and welfare coverage. It wasn't as if he had not earned health coverage. It was the rules that governed the pension plan that decided if he would get health coverage or not.

Joe asked me to give him a few days and return. When I did, he told me he had contacted United States Lines told them about my father's situation and they agreed to hire him back for six months.

Joe had never met my father. He could have listened to me and told me there was nothing he could do. Instead he had done what he was sworn to do, help out a fellow union member in time of need. To my knowledge this was the only time my father needed help and he got it no questions asked.

I was hired directly by District 190 and went to work. I was starting over, at the bottom; once again I was an outsider. But I was going to pay back Joe Bobo for his kindness to my family, and not embarrass Mike Day or Don Crosatto.

Organizing for the Machinist Union was different. All their organizing drives are bottom up. That means using the National Labor Relations Board, which requires a Labor Board Election supervised by the Federal Government. At the carpenters we used the labor board election process in the Industrial Unions but rarely for Construction Organizing. I had done bottom up organizing, but to be honest was a little rusty.

The Labor Board process is unfair and is weighted in favor of the employer. Republican administrations have weakened the process so far in the employers' favor that unions can seldom reach a negotiated collective bargaining agreement. Democrats seldom support legislation forwarded to them by the labor movement. And they seldom fix the inequities their Republican brothers have left in their wake.

The Labor Board is rife with representation that either favors the employer or is incompetent. I know of statements taken by drunk Labor Board attorneys, decisions against a worker without interviewing the worker who filed the unfair labor practice charges, decisions rendered in favor of the employer without interviewing or investigating the employer, or a Labor Board hearing officer who ruled in favor of the company and allowed, the company to stop bargaining, dismissed our charges because the case would postpone his retirement.

The Labor Board Process is also unfair to workers and the unions forced into using the government because in many cases employers hire consultants who specialize in breaking unions, who illegally demoralize the union work force by using fear and intimidation to beat the union election.

These consultants are the lowest form of human being on earth. They make their living ruining the lives of decent human beings who are tired of mistreatment. They decide who stays and who is fired, they pit worker against worker, rewarding those who turn on a friend and punishing those who stand up for themselves.

In America, where freedom and democracy are praised and defended as the most important value we hold dear, it is surrendered that moment a workers crosses into the plant or job site. For the length of their shift every American gives up their right to free speech, the right to decide if he must choose to work in a dangerous situation, as well as the right to know if the substances they are handling are toxic. They are forced to attend meetings that are irrelevant to their work. They are prisoners, who without unions are forced to accept dangerous and inhuman conditions eight hours or more a day, just so their families can eat.

It sickens me that in the land of opportunity, the country that demands democratic and fair elections in foreign countries, the same evil forces that wrote the legislation creating the National Labor Relations Board in the first place allow federally supervised elections in the United States to be rigged in favor of the employers.

In the first phase of an NLRB election, the union must present a petition to the government that shows 51% of the workers signed to establish representation. The company usually finds out that the workers are organizing so they often begin to retaliate against them. Then the petition is presented to the company and the company has the opportunity to contest the conditions of the upcoming election or eligibility of the voters. In many cases the company tries to make supervisors eligible to vote. Any consultant can foul up this process for a few extra weeks while the company is allowed extra time to beat up on the work force.

When the election date is set, usually six to eight weeks out, the company has a series of "captive audience" meetings that the workers are made to attend on company time. The union is never allowed to attend these meetings. The union must hold meetings off site and they are certainly not mandatory.

By Election Day most workers are overwhelmed, intimidated and feeling as though they are in a lose-lose situation. And then if they vote yes, and the union wins, which is less than 50% of the time, the company can legally stall the bargaining process, and continue the mistreatment of the work force.

Sure, the union can file charges against the employer with the NLRB, but what do you get from that, a board agent who is more interested in moving paper from one pile to another? Who will ultimately give up on the workers' rights in favor of a closed case file? Or the case will slowly wind its way through the black hole of the NLRB's decision-making process, taking months and years, only to find that there was some loophole that allowed the company to continue breaking the law. Or after six years, one case in a million is ruled on in favor of the union and by then all the workers have quit or been fired. Then the union must negotiate the contract.

So, if you want a union you must vote up to three times, once by signing the petition, once in the election and then a third time again on the contract if the union can get one with the employer. The process is unfair and bullshit. So how could I overcome this, how could I win? I learned to cheat.

My first effort for the Machinists was against a company called LTD Ceramics. LTD employed about 170 machinists. They manufactured high quality, ceramic machine parts for the computer industry. The parts were used under a patented process where after being machined they would be exposed to intense heat.

The company was run by people who I came to believe were some of the biggest assholes on earth. The work force was American 10%, Filipino 30%, Hispanic 10%, Vietnamese 30%, and 20% Romanian immigrants. The reason the workers had come to the machinists union was because they were being treated unfairly.

They were forced to buy their own tools, while making less than $10.00 per hour. A simple tool kit could cost as much as three hundred dollars. The company would buy them for the workers and deduct the amount from the workers first paychecks. Many workers were terminated before they had paid for their first set of tools.

Cameras were set up everywhere so that the company could watch every step a worker made. There were complaints that overtime was calculated differently for each racial group. Women were harassed by their supervisors. Workers were cheated out of time off.

Of all the workers in the plant the Filipino workers were the most fearless. At captive audience meetings they would stand and defy the boss, they would argue with the consultant and they would work harder than anyone else while doing their jobs. They built the committee, moving the petition throughout the plant, recruiting new people from the other nationalities.

I was always alone except for the workers and the committee. District 190 could afford to pay me, but they expected me to do it by myself. I had to trust in the committee to carry their load.

As the company committed Unfair Labor Practices, I filed them. I drove the workers to NLRB to give statements. I covered LTD with paper, charge after charge. I hand billed the work force twice a week, standing on a wall that was in front of the company. The cops would come and I would refuse to leave, waving my Union Yes Banner in their faces. I would insult every supervisor who mistreated a worker in front of the workers as they left for home. And I house called every worker at home as many times as I could, and took the committee with me whenever there was a language problem.

I taught the committee how to organize and how to anticipate the company's next move. We stole the union buster's manual and inoculated the workers with his next move before he could. The committee knew the union's response to issues before the consultant asked the questions.

Our only problem was reaching the Romanians and the Vietnamese. While they understood some English, I could not breach the language barrier. I was disappointed but optimistic; my information had us winning by 20 votes. I considered withdrawing the election, but I knew my committee was dead meat. I told them the situation and they didn't care about losing their jobs they just wanted to win. Against my better judgment we went forward to the election.

On Election Day we won by ten votes. It seems the unions were very strong in Romania and many of the Romanian workers had been union members back home. Mike Day and Don Crosatto were thrilled. This was the first machine shop and the biggest bargaining unit District 190 had won in years.

We won because the committee was so strong. I gained tremendous respect and admiration for the Filipinos in the plant and on the committee. And because I wasn't afraid of them running their union, they were union long before the election took place, they were union in their hearts.

Still, the worst was yet to come following the election. The company laid off 50% of the work force. We spent weeks negotiating the conditions of the lay off. Negotiating Committee members were terminated. We filed more Unfair Labor Practices, to no avail.

On our side, Mike Day assigned the business agents for the manufacturing local to negotiate the agreement. They were good guys, but used to negotiating agreements that had existed for decades were based on long-standing relationships with owners. The LTD negotiations were with someone who did not want the union, someone who would agree to conditions begrudgingly at best. This was a first agreement and could not compare to the existing agreements. In their minds they could not agree to new conditions that would have to be improved upon in each consecutive agreement.

I was going mad, the committee was getting mad. It was an organizer's worst nightmare. The company was giving in to limited conditions and dragging out the clock and the union could not accept conditions that were less than their existing master agreement. The company finally fired or laid off enough workers so that our supporters were gone. The supervisors illegally led a de-certification. The neutral arbitrator announced his retirement and ruled in favor of the company to clear his caseload and in the end I turned the company in to its main customer Applied Materials and the company was caught by their largest client counterfeiting the parts they made.

Working for a good boss like Mike Day allowed me to create an organizing department built around the organizer. In an organizing campaign there is the union organizer and then there should be the real union organizers, the workers.

The Union Organizer has to develop leadership in the workers, direct the campaign, and prepare the workers for everything the employer is about to throw at them. He or she must train the workers to take control of the organizing effort inside the workplace. He must be willing to allow the workers to win the campaign.

Let me give an example of what I mean buy "allowing the workers to run the campaign". I received a call from a group of mechanics at Fremont Pontiac, who wanted to organize. I began inoculating them on what to expect. After signing the NLRB Petition they delivered it to the boss and demanded recognition. When the boss refused, the company hired a consultant to beat the union, the men verbally attacked the consultant, they were prepared to counter every antiunion message the consultant delivered. The company gave up on captive audience meetings.

On top of the house calling I would arrived at the shop at 7:00 am every morning and campaigned inside the premises until 7:45 am when the supervisor would arrive and I would sneak out the back door. I used this method in every shop that I could sneak into. For six weeks I was on the shop floor every day and we won with 100% voting for the union.

Then I went to Mike Day and told him I needed a negotiator who was not going to demand conditions in the agreement that had been negotiated in the older agreements over 20 years. What I mean is District 190 was an old established union; many of their agreements where 20 to 30 years old. In those agreements were conditions the members had earned over the course of many years. In these newly organized shops some of these conditions were unreasonable. And more important is that some of these conditions were unimportant to the men who had just fought to organize the union.

In addition I asked Mike to make sure that the organizer who had fought alongside the men was part of the organizing committee. The organizer has the relationship established with the work force. My own experience with both unions proved disastrous when the union would win an election and then tell the organizer to wait outside. By the time a union wins an election the employer should be afraid of his workers.

Fortunately for me after winning an election Don Crossato and I would negotiate the first agreement, or I would negotiate with the organizer who had been involved.

And last but most important is the workers must be at the bargaining table, sitting with the organizer and negotiators they know and support.

Jesse Jaurez and Pedro Mendez were both union organizers who I worked with at District 190. They are both excellent organizers who know how to win and have.

When I first went to work for the machinists Jesse was either winning elections by one vote or losing them by one vote. I suggested that he involve the worker more in the effort. He saw how I was doing it and he began to use the same methods. His win rate flipped, he is one of the best organizers I know; he wins elections and negotiates first agreements.

Pedro Mendez was the same, barely winning elections or losing them outright. We were assigned to run a campaign against Swissport, a company that handles baggage, cleaning and miscellaneous ground duties at the San Francisco Airport. The City of San Francisco hand passed an ordinance that workers must be paid a minimum wage of $11.50. This ordinance also required those non-union companies working on Airport property agree to card check provisions if their employees choose to organize. This meant the unions did not have to use the NLRB process. If a union notified an employer that they had 51% of their employees signed to an authorization card they had to recognize the union and begin negotiations.

Now this sounds easy--get the cards signed, notify the company, prove majority status and negotiate the contract. Not so easy, the companies still fight the union. They still spy on workers, hire consultants, and lay off or fire as many union supporters as they can. So anyone who believes that "Card Check" is the answer to America's organizing efforts will learn they are still facing the same obstacles faced in the current climate.

The workers' committee would sign any new hire immediately or Pedro and I would find them somewhere on Airport land or house call them at home. We had reached a point where on any given day, we would have 51% of the authorization cards from the 270 person unit. Then Swissport would lay off a group of workers, hire a new group of workers and we would start all over. I made a decision that we would have to file the cards with the neutral arbitrator and hope for the best.

Fortunately for us the arbitrator was fair and impartial; I pulled one of my best tricks ever during the card count. We needed one more vote than 50% which was 92 yes votes. Pedro and I could count 93 yes votes, far too close to call for the card check. But the company was firing and hiring as quick as we could get the new employees signed up. So we decided to call for the vote. Gerry McKay, a very well respected labor arbitrator, was named to conduct the card count. He had the agreed-upon list of names and threw everyone out of the room except a representative from the machinists international union and myself.

For some unknown and unexplainable reason I can read upside down as well as I read right side up. We had a stack of cards four inches thick. Many were doubles, cards signed by workers more than once. The stack also included all the cards signed by workers who had quit before the count. I did this to intimidate the Swissport management.

As I read the names in alphabetical order to the arbitrator he would check them off his list. If the next card was a double, I would say the name of the person who was next on the list, but not signed to a card. The arbitrator would check the name off. I knew the list almost verbatim so I added on about 10 more names and by the time we were done we had won the election by a far greater amount than the real number of cards we had signed.

We elected our negotiating committee from among the workers and reached an agreement.

During my time at District 190 I learned a lot about how other unions work; many of these things I carried with me for the rest of my career as a union organizer. The most powerful lesson is that workers must be involved in the fight and at the bargaining table. Workers are the union, when workers believe that they must become part of the effort and take personal responsibility for the outcome, success is at hand. When they put their job on the line and stand up to the boss everything is possible.

When the organizer does all the work and fears the workers as much as he fears the employer he will most often lose. When the union fears the workers they have a relationship with, the unit will be lost at the bargaining table. The union exists in the hearts of workers long before they have a contract or official union membership.

Chapter Twenty-Four: *Local 34 Fights Back*

Local 34 is an interesting local union, while this chapter may be a little long, I think it may explain why pile drivers are an independent group and why the members actually believe that they own it. In all my years I saw and learned democracy from the jobsite to the floor of the union meetings. I will draw two interesting pictures of "pilebutt democracy".

I was working on a job driving concrete sheet piles at the Oyster Point Marina about 1979 The superintendent decided that it would be cheaper to have the laborers place the hangers on the sheets, rather than pile drivers. Red O'Niel, a crewmember, told the foreman Ralph Freeman that the work bolting the hangers on the sheets was pile driver work. You also have to know that these two men were both over six foot two, and a good three hundred pounds each.

Without blinking an eye Ralph Freeman, told the superintendent that if the laborers continued doing our work we would not work. The laborers continued bolting and to a man we sat down and stopped working. We did not wait for a business agent to come and resolve the problem, we stopped working. Within minutes the laborers were taken off the work and we went back to work--On-the-job democracy lesson number one.

Lesson number two is even better. In the early 80's while I was president of local 34, an intoxicated member made a motion that was completely unreasonable and ridiculous. I ruled the motion out of order and refused to allow the motion to be seconded. I was just trying to move the meeting along a little faster. Other members jumped to their feet and challenged the chair's ruling. I was verbally attacked for being a tyrant and undemocratic.

Under Robert's Rules of Order when a motion is challenged from the floor the vice -president takes the chair and both parties argue their position. I argued that the motion was unreasonable and ridiculous, the member stated his position, which in my opinion was even more unreasonable, and ridiculous. A motion was made to accept my argument or the member's argument. The membership voted that as president I had to accept the member's motion.

I then placed upon the floor the motion made by the member which was seconded, I called for the vote and the motion was voted down. My lesson in democracy was that no matter how ridiculous or irrelevant a member's ideas the membership would decide if it was valid or not.

These two lessons best describe for me what it is to be a pile driver. Long before this fight I was educated on the job and at union meetings that there was never a choice to make between the Carpenters Union and Local 34; that there is no choice when picket lines goes up. Being a pilebutt is painful, and without the union many more of us would die or be maimed. Missing a day's work or walking off the job when things go too far is expected. The loss in pay is not considered or moaned about. The loss of face or having a bad name and reputation is a greater concern and can end careers. We are a tightly knit community of men and women and to be shunned by your brothers and sisters whose respect you need is the best lesson in unionism I have ever learned.

Even the fact that many of the employers are card-carrying union members creates a unique understanding between employers and employees that I have never seen anywhere else. A good crew makes employers lots of money, their knowledge of one another, how they respond to each other and even how well they work as a team, makes our local 34 unique. An owner or superintendent who has lugged around the same sorry chokers as you have better understands your situation than some sorry college students who don't have a clue. Don't misunderstand this relationship; it is still worker versus employer but there is a level of mutual respect. Someone has to do the job and we do it, we work together, die together and fight together.

Pile Drivers Local 34 has always had a strained relationship with the Carpenters Union since they were forced into the UBC in 1920. I hope to explain the development of the local union's ideology and perceptions to anyone who is not a pile driver, and also the development of the mystique which surrounds those of us who lay claim to the title "pilebutt".

The Pile Drivers Association founded in 1849 became the San Francisco Wharf Builders Union Local No. 1. They had been an independent union established in 1849 in San Francisco.

1901 was a year of great change for the Wharf Builders Union No. 1. On May 1, 1901 the Wharf Builders Union received a Federal Charter from the American Federation of Labor and became Pile Drivers Local Union No. 9078.

The members who signed that charter were T.J Denahy, J. Hodgson, W.W. Carthy, Thomas Cosgrove, L.W. Mullen, R.F. Cook, J.V. Beck and J.S Steward; for the American Federation of Labor Samuel Gompers signed.

Pile Drivers and Bridge Builders Union Local 9078 received its charter from the AFL as a result of the Federation's attempts to bring the many independent unions under the umbrella of one national organization. This was no easy task if you consider the growth of unions nationally and the many local organizations that had tried and failed to form structure and order among the varied trades that were being created.

The American Federation of Labor under Samuel Gompers along with the leadership of the International Unions began the transition that would bring the many independent unions under the umbrellas of the varied international unions and along with them came the battles over work jurisdiction.

On May 21, 1904: A motion was made that the Pile Drivers and Bridge Builders Union 9078's officers be declared the regular officers of Pile Drivers, Bridge and Structural Iron Workers Union No.77 was carried. The step over from an independent local union was completed. By affiliating with the A.F.L. the Pile Drivers Association of the 1870's was now part of mainstream unionism. The plan of the AFL was completed as far as the Pile Drivers were concerned. The minutes of the time show no unhappiness on the part of the membership in becoming part of the International Association of Bridge and Structural Iron Workers.

Following this affiliation the International Unions of the newly formed AFL began fighting with one another over which union had jurisdiction over the work on the job sites.

On October 1, 1910, the Los Angeles Times Building burst into flames, a suit case filled with dynamite exploded killing twenty people and caused a half million dollars in damage. The man hired to track down those guilty of the bombings was William J. Burns of the Burns Detective Agency. It was his opinion that the Iron Workers Union was responsible for the bombing. The Metal Trades of Los Angeles were on strike when the Times Building was bombed. The Iron Workers, being the strongest union at the time, made them a prime target.

International Secretary-Treasurer, John J. McNamara, was arrested in Indianapolis for conspiracy along with his brother James McNamara, an active member in the Painters Union. The Iron workers were defended by Clarence Darrow, the famous trial lawyer. The trail ended when John McNamara accepted a plea bargain so as to save the life of his brother who had been arrested for murder.

McNamara was sentenced to 15 years in prison. In 1916, two men were arrested and confessed to the bombings of the Los Angeles Times Building and were no way connected to the Iron Workers Union. It also appeared that anything of any value to the newspaper had been removed from the building. The Times never lost a day of work.

President Taft ordered the Attorney General to investigate the bombings and fifty men were arrested for illegally transporting explosives – mostly iron workers, but some were carpenters, steamfitters and machinists. On June 25, 1914, fourteen Iron Workers presented themselves at Leavenworth Prison to begin serving their sentences.

During this same time period a jurisdictional dispute broke out between the Iron Workers Union, Carpenters Union and the Lathers Union over work being performed by the Iron Workers in New York City.

Coinciding with the convictions of the Iron Workers in 1914 the New York State Federation of Labor awarded the Independent Dock Union Charter to the Carpenters International Union who then created Carpenters Union Local 1546. The AFL then revoked the charter of the Municipal Dock Builders Union who affiliated with the Iron Workers International Union setting the stage for a huge battle over jurisdiction of all Dock Building in New York City. The Carpenters union and Robert Brindell then pressured the employers one by one to sign agreements with Local 1456. The relationship between the Iron Workers Union and the AFL continued to decline until the Iron Workers were expelled in 1916.

Robert P. Brindell joined the Independent Dock Union in 1905. The Independent Dock Builders Union (IDU) like its sister local in San Francisco received its Federal Charter from the A.F.L in 1907. The IDU lost its Charter in 1910 for non payment of per capita taxes and within a few months a new dock builders union was formed as the Municipal Dock Builders Union was chartered by the AFL, creating two dock builder local unions in New York City.

General President Hutcheson then moved to take control of the New York City District Council of Carpenters and suspended all sixty-three local unions, revoking their charters. Through Brindell's violent acts against the membership and force against the employers the Carpenters International Union successfully consolidated the New York District Council of Carpenters into 1,000-man local unions and appointed of each local union president.

Brindell was rewarded with the leadership of the New York Building Trades. He gave himself a monthly salary of fifty cents per member making him the highest paid union official in the United States. He would be found guilty of extortion and sentenced to five to ten years in Sing Sing in 1921. He was released in 1924 and died in 1927.

Hutcheson used Brindell to take over and consolidate the New York City District Council of Carpenters. The successful takeover of the New York Dock Builders resulted in the jurisdictional recognition of all Pile Driver locals by the AFL. And by 1917 the Iron Workers dropped "and Pile Drivers" from their name and were reinstated into the AFL. From 1917 through 1920 Pile Driver Locals were affiliated with the Carpenters International Union, they were issued new charters and in May of 1920, Pile Drivers Bridge Wharf and Dock Builders Local 34 received its Carpenters Union Charter.

By 1910 the Iron Workers began to lose power, not by choice but by circumstances: The bombing of the Los Angeles Times building, the internal fight with the AFL, the fight over the Pile Drivers Jurisdiction and a fight over the jurisdiction with the Lathers over the installation of reinforced iron. The weights of these many action at the same time forced to Iron Workers Union to let the Pile Drivers go.

1926 was a period of revolution for unions in the San Francisco Bay Area. The San Francisco Building Trades had negotiated an agreement with the employers without ratification of the membership. Building Trades members went on strike without the permission of the Building Trades or the International Unions. They formed the Rank and File Committee and organized to fight the employers but also fought openly against the unions they perceived to have sold them out and as a result Pile Drivers Local 34 and Carpenter Locals had their charters revoked by the International Union.

The AFL represented the tradesmen, Carpenters, Iron Workers, Plumbers, Electricians , etc; men who considered themselves craftsmen and as professional as any doctor or lawyer, the elite of the working class.

The Congress of Industrial Organizations (CIO) was a new organization representing those industrial union members, Longshoreman, Mineworkers, Auto Workers and any union that may not fall into the classification of tradesman defined by the AFL.

The CIO was new and progressive, socialistic, while the AFL was conservative and paternal. The Pile Driver union was split into factions: those loyal to the ideals of trade unionism and those loyal to the socialistic ideals of the CIO. In addition the Pile Drivers work was predominantly on the water building docks, piers and bridges. The union had always associated with the other waterfront unions like the Longshoreman and the Sailors Union of the Pacific belonging to the waterfront federations and following CIO ideology rather than the more conservative building trades' philosophy.

The strike that resulted was lost and San Francisco operated under the American Plan very much like the Merit System of the 1980s, the result was nonunion men working along side union men. This was also a time of social and political unrest.

But for Local 34 the American Plan had little effect, the pile driving crews were by and large small and related through blood, race and ethnicity. Company owners were members of the union, crews were made up of brothers, cousins or fellow countrymen. The work got done and union rates were paid regardless of the strike.

The union continued to exist with or without a charter from the International Union. A second Pile Drivers Union was formed, work was dispatched and meetings held by both unions until a truce was negotiated and the workers reunited under one union. But a line had been drawn in the sand by the International Union and a distrust of leadership who could act without vote of the membership was created when the Building Trades and the Carpenters International Union acted without direct membership participation.

Prior to 1936 the Carpenters International Union had organized and lost many of the lumber and sawmills of the Pacific Northwest. At one of the general conventions the entire delegation of lumber and sawmill workers walked out of the convention in protest and formed their own union, which affiliated with the Congress of Industrial Unions (CIO).

Many of Local 34's members had roots in the timber industry and with the Industrial Workers of the World (IWW). The Saw Mill Workers (CIO) and former UBC members were organizing saw mills, on strike against their employers and fighting the attempted raids of the UBC. They had asked the Central Labor Councils throughout California for a five-cent per member tax to aid them in their fight. Local 34 voted to support the sawmill workers and pay the per capita tax.

William Hutcheson, the Carpenter's International president, put local 34 under trusteeship and appointed a temporary executive board to replace the elected executive board. The members revolted and stormed the union hall on Bryant St. in San Francisco. The members lived in the union hall and controlled all dispatching of work in Northern California. The International expelled the elected Executive Board, and replaced them with anyone who would volunteer, but the membership stood strong for their elected representatives.

In one case Hutcheson sent his cronie Joe O'Sullivan out on one of the piers on the Embarcadero. His orders were to kick the anti-Hutcheson Pilebutts off the job and man the work with carpenters. O'Sullivan and his fellow carpenters were beaten and sent on their way. The end result was a truce, local 34 agreed not to send the CIO strikers any money and the International agreed to leave local 34 alone. This fragile truce was kept for almost sixty years. Local 34 maintained its autonomy, elected their business agents and negotiated their collective bargaining agreement without help from the UBC.

I have always taken pride in that this story had been retold to me by Jack Wagner the leader of the anti-Hutcheson Pilebutts, and I retold it in my first book. By the time the International was about to take Local 34 over for the second time, the membership and leadership knew that if it had been done once, someday it would be done again.

I was taught by Pilebutts and every member that has been also initiated into local 34 knows that if it happened twice it would happen again. Like most pile drivers I have been taught that the union is mine, that the fight is mine and I have a responsibility to stand up and take a side.

Chapter Twenty-Five: *Pile Driver Pride*

On May 27, 1997, the newly Northern California Carpenters Regional Council (NCCRC) was established, and all the Carpenters Business Agents and Local 34s elected business agents were informed that they were now on the payroll of the NCCRC. They were required to resign their positions and were rehired as Field Representatives. Within days, Local 34 was informed that the contract would be negotiated by the NCCRC, and that no rank and file members would be part of those negotiations which was a longstanding practice at the Local. Also, that the members would not vote on the contract and that Local 34 could no longer dispatch pile drivers to the bridges being built all over Northern California. On Monday of the following week I quit and resigned my position as President of the Regional Council. I could not come to terms with the buy-out of full time union officials, having the opportunity to take pension credits for work they would not do. And what I was going to be expected to do regarding Local 34.

Then on July 8, 1997, the Carpenters International Union held a hearing to determine if Local 34 should be put into trusteeship, which in effect would allow the International Union to take control of Local 34, including the property, treasury, finances, dispatching of jobs and operation of the union. Representing the International Union was Fred Carter, General Executive Board Member, Sixth District, Harry Gowan, General Executive Board Member, Third District and Bill Halbert, UBC Representative, Second District. Also in attendance were Michael Draper, General Executive Board Member, Eighth District, John Steffens, District Representative and John Casey, Interim Secretary Treasurer, Northern California Carpenters Regional Council.

The Executive Board of Local 34 was accused of refusing to affiliate with the NCCRC. The elected officers of Local 34 had refused to submit the required forms for employment at the NCCRC and the union had been closed on two specific days with-out contacting the NCCRC, theoretically depriving the membership of work opportunity.

The outcome of the meeting was pre-determined and a formality. Local 34 did attempt to get a restraining order against the UBC but the effort failed. On July 30, 1997, General Executive Board Member Mike Draper and about twenty Carpenters Union Representatives attempted to forcefully take over the union's hall on Hegenberger Place in Oakland.

Fortunately, Senior Business Agent David Rhodes was in the building and locked the gates that surrounded the Local with him inside. Draper informed him that he was there to take control of the local. Rhodes wrapped his arms through the gates covering the locks and told Draper he would have to break his arms.

The Oakland Police were called and the police officer asked to see the paper work that allowed the UBC to take control of the union and its assets. Draper had no such papers and the crowd of Carpenter Business Agents retreated back to the Regional Council Offices.

By September of 1997, though, Local 34 had exhausted its efforts to hold out the International Union and the Trusteeship was finally implemented. While I had already quit around June 1, 1997, I was soon followed by the elected leadership of Local 34; to their credit they were David Rhodes -- Senior Business Agent, Thom Donnelly – Financial Secretary, James Walker – Business Agent, and Larry Wagner – Business Agent. The Executive Board was dissolved and for the next eighteen months Local 34 operated with an Executive Board of Union Meetings. Carpenters business agents were brought in to police the Pile Driver's work and to run the hiring hall until the Regional Council could find pile drivers who would work for the union.

It is at this point where the importance and power of shunning must be noted. Local 34 has a character about them. If the crew determines that action must be taken then the majority rules, democracy in action. It took the regional council four months to find a Local 34 member to cross the street and enter the union as an unelected representative. One individual came back from retirement only to be insulted, disrespected and shunned by the membership. Another left a comfortable union position only to receive similar treatment at the hands of the membership.

Because the Regional Council refused to hold union meetings, the unseated Executive Board and others formed an alternative organization called Pile Driver

Pride. Their logic was if the NCCRC would not allow the members of Local 34 to meet, then they would hold their own union meetings and form a parallel organization to shadow the existing union. A Pile Driver Pride Executive Board was temporality put together. They were Phil Daniels, Thom Donnelly, Doug Downie, Pat Karinen, Bruce Lanham, Robert Lanham, James Lawson, Dan Murry, Red Poulsen, David Rhodes, Bob Smith, Steve Tilton, Bert Tycer, Larry Wagner and Jim Walker. They were all elected unanimously at the Pile Driver Pride meeting held at the Longshoreman's Hall on February 7, 1998. Pile Driver Pride bylaws were voted on and amended and lawsuits were filed on the membership's behalf. For the next eighteen months Pile Driver Pride held union meetings in place of the canceled regular union meeting. Their founding principles were Rank & File Democracy – One Member One Vote – rank & file ratification of all contracts including modification and rank & file vote on and approval of all Local Union and Regional Council Bylaws and UBC Constitution.

A monthly newsletter was published and funds were collected to maintain the organization. Pile Driver Pride crews continued to elect their stewards on the job sites and police the pile driver work independently of the local union. But most importantly 99% of the membership refused to cooperate with the appointed union representatives.

In 1999, after more than a year, a union meeting was held and elections were called for. The Regional Council found a few members who would run for office. But on Election Day the entire Pile Driver Pride Slate of candidates were elected into office with David Rhodes as President of Local 34. Unfortunately, much of the pile driver work had been taken over by the carpenters locals and work assignments, always done by pile drivers, was allowed to be done by other trades due to lack of enforcement by the carpenters agents. Since the Pile Driver Pride slate of officers had been swept into office by the membership Pile Driver Pride became redundant and slowly faded away. Union meetings were tough; the carpenter appointees were shunned and belittled in the union hall and in the field. Regardless of the Regional Council control, the union meetings and job stewards were controlled by Pile Driver Pride.

Our next goal was to retake the Regional Council Executive Secretary position.

CHRISTIE

Robert Christie Jr., a lifelong resident of the San Francisco Bay Area, passed away suddenly at his home on January 20, 2000 at the age of 50. He follows his beloved wife Ewa Christie, who he married in 1968. Devoted father of seven daughters, Julia, Isabel, Rose, Delores, Anna, Catalina, Veda and one son Robert. Fifteen grand-children and two great-grand-children survive him.

He was drafted into the United States Marine Corp in 1969 and served in Vietnam where he was seriously wounded, and was decorated with the nations second and third highest honors.

He joined Carpenters Local 22 during a "Wildcat Strike" in 1974 when he left his non-union job to join the carpenters union and the strike. He was a member of Carpenters Local 22 until his premature death. Over the past year he was an advocate for change and democratic reform.

He graduated from the University of California in 1984 with a degree in Philosophy but continued to work with his hands as a carpenter. He took great pride in pointing out the various buildings he had worked on. He enjoyed telling stories to his grandchildren who will miss him dearly.

At the time of his death he was surrounded by his family and passed peacefully into the arms of his fallen comrades.

The family will be holding a private memorial service. His ashes will be combined with those of his wife Ewa and will be scattered along the Pacific Coast line by the Neptune Society.

Courtesy – Hayward Dailey Review

Chapter Twenty-Six: *The Day I Killed Robert Christie Jr.*

It was about eight p.m. one night when I received a call from a guy who introduced himself as Ken Little, the leader of Carpenters for a Democratic Union (CDU). Little had created CDU as an organization to combat the United Brotherhood of Carpenters' undemocratic actions, with him as the leader. He was trying to build it into a viable organization of members within the UBC

He had been told that I was Robert Christie Jr., the guy on the Internet writing about union democracy. He stated as a matter of fact that he was obligated to "out me" for the good of the union and that I was needed to run for General President of the United Brotherhood of Carpenters and Joiners. We spoke briefly, but I never revealed to him that I actually was Robert Christie Jr. I had never met Ken Little, and was amazed that he called himself the leader of the Carpenters For a Democratic Union, since no election was ever held to elect him to that office. And one of the premises of our fight was election of officers by the members.

The next day, I killed Robert Christie Jr. Early in the fight against the takeover of Pile Drivers Local 34 by the newly formed Northern California Carpenters Regional Council, I had decided to focus my efforts on building a stronger dissident movement within the UBC. My motives were to create another diversion for the UBC to focus on, in the hope that they would spend less time on Local 34.

One night I was staring up at my book collection and thinking about a name to use on the web. Staring right back at me was the book "Empire in Wood" by Dr.Robert Christie. "Empire in Wood" is probably the most critical and honest history of the United Brotherhood of Carpenters. It also spells out the McCarron plan to return the UBC to the Hutcheson era of conservative unionism. After reading it again I saw in it the blueprint for the restructuring of the Carpenters Union. As if struck by lightning, I decided my web persona would be Robert Christie Jr., grandson of Robert Christie, author of "Empire in Wood", and member in good standing of Carpenters Local 22 San Francisco California. I began to write and fight. I wanted to participate but I felt if I came out and spoke my opinion I would have become the center of the controversy and that the questions which needed to be answered would be deflected and left unanswered. I also knew from organizing workers how to start a fight with the boss, how to pick at the wounds of lies and deceit and keep them bleeding, how to expose dishonesty and educate workers and introduce the concept of unionism into their mind until the ideas become their own. And then turn them loose on the boss and to create their own personal organizing campaign. I thought that by using a pseudonym I could address change. And to be honest my wife Leslie had also paid a price in this fight and I didn't want her worrying about me losing my job as Director of Organizing with the Machinists Union. And I didn't want the Machinists Union to become embroiled in a fight not of their making. However Don Crosatto did know I was still involved to some degree and allowed me to continue in my efforts.

The real Dr. Robert A. Christie, the grandson and son of union carpenters, worked his way through Swarthmore College as a union apprentice. He originally wrote "Empire in Wood" as his doctorate dissertation while attending Cornel University in 1954. Although one of the great labor history books, it was not funded by a labor union and was written by an historian without propaganda or paid for with union funds. In response to Dr. Christie's book the United Brotherhood of Carpenters commissioned Maxwell C. Raddock to write Portrait of an American Labor Leader – William L. Hutcheson. Hutcheson, being the General President of the UBC, paid Raddock $310,000 to respond to the critical "Empire in Wood".

In sworn testimony before the, "Select Committee on Improper Activities in the Labor or Management Field", Dr. Christie testified and submitted evidence that Raddock had plagiarized "Empire in Wood" and submitted thirty-three pages of Dr. Christie's writings that had been transferred into Raddock's official history of the United Brotherhood word for word as well has the writings of many other respected labor historians. During the hearings it was also uncovered that Raddock had paid a private detective to check on information about AFL-CIO President George Meany, in the hope that the information could be used to "stiffen his back" in dealing with Walter P. Reuther, president of the United Auto Workers Union.

Dr. Robert Christie died in a plane crash at the age of 46 with his wife Sally, and two children, Ellin, 12 and Thomas, 9 while flying from San Pedro Island to Belize, Central America, gathering archeological data for his wife Sally's doctoral dissertation.

I began writing, trying to develop a platform and create some kind of dialogue regarding the direction that the members who wanted change in the UBC could take, as well as to turn attention to the NCCRC Executive Secretary John Casey and his failures. I intended to help get Bob Alvarado elected as Executive Secretary Treasurer at the next election. I posted a series of articles about union democracy, self determination and even wrote a resolution that reached the floor of the UBC General Convention. The "Christie Resolution" as it was called was passed on the floor of enough local unions to have it voted on by the delegates at the 2000 General Convention.

At the time, I was Director of Organizing for the Machinists District 190. I was overseeing the organizing activities of the International Association of Machinists, running organizing campaigns myself and then working at night being Robert Christie Jr. So, when Ken Little called I realized that maybe Robert Christie Jr. had overstayed his welcome. The following day I went to the Hayward Daily Review and submitted an Obituary on behalf of the family of Robert Christie Jr. and his "untimely death", knowing of course that everything about Robert Christie Jr. was fiction and a fabrication developed to divert attention from Local 34. For weeks he was mourned by rank and file carpenters through their postings on the Internet, but his untimely death helped propel the resolution to the floor of the general convention. This may go down in history as my greatest prank ever…except maybe the cartoons.

As I began to mentally reconstruct myself after quitting the McCarron/Casey team, I began following the carpenters unions on the Internet. What an amazing and revolutionary machine the computer can be…with the Internet, there are few secrets; once someone knows something and posts it out there for all interested parties to see, everyone knows it. Anything is said, truths, lies, fabrications, personal attacks and once in a while deep and revealing thoughts. All around North America union carpenters were creating various forums to speak out and discuss the UBC. Jawin, UBC One List, Carpentersunionbc are the oldest among the dissident web sites I'd peruse. I watched and read, the sites displayed little direction but isolated pockets of union members were beginning to speak out. The World Wide Web is the breeding ground of revolution and anarchy. And while I was mad as hell about what had happened in Northern California, I did not want to see the membership and contracts that have taken over one hundred years to achieve, be destroyed. The constant barrage of negative remarks and destructive personal attacks that take place in these forums is amazing and the reason no dissident movement within the UBC or any labor union can hope of taking power.

"UBC Onelist" was one website supported by Yahoo Mail. People post messages and they are mailed out daily to everyone who has signed up. Typically, they are angry messages about the UBC, telling others how awful the Carpenters Union is and urging others to revolt against the leadership. It's very frustrating because I believe the Carpenters Union is the members and that tearing the union down without a clear program to rebuild it could be devastating to the men and women who belong or want to the Union.

"JAWIN" was another web site that was more like a bulletin board where people post messages and others may respond or post something different. This site was open to all and is one of the first and oldest sites. It is like Onelist but much more brutal, the one advantage to Jawin is that it was totally anonymous. Those running that site have kept the names of those participating private. This site is also the one site constantly attacked by computer hackers. It has been somehow knocked off the Web on numerous occasions, but it always bounced back on the air as belligerent as ever. It is also a great source for members who are interested in what is happening around the country in the UBC. It is total anarchy and anything goes; no one is safe, anti-McCarron or pro-McCarron, no one is safe.

And then there is "carpentersunionbc", the carpenters of British Columbia who have seceded from the UBC and merged into a Canadian Union. This is the most legitimate dissident union website, full of useful information posted with an ample dose of humor, and is funnier than hell. They post timely information about the Carpenters Unions around north America..

As the NCCRC election approached I worked less on matters outside northern California and began to focus on the upcoming Executive Secretary Treasurer election and getting Bob Alvarado elected. During the 1991 convention in Atlantic City I met John Liguori, a member, I believe, of one of the New Jersey locals and a friend of Tom Pearl, a Whitey Rogers supporter from Mountain View, California. I don't believe John was a delegate to the 1991convention. But he was a union carpenter who wanted the UBC to change and he came on his own to change it. When Tom introduced us, he said John was a good guy and longtime friend.

In 1999 when I first went online, John Liguori was still fighting to change the union, posting messages and arguing for a different UBC. We began to correspond and our friendship has grown over the years. I posted alongside John during this time as "Robert Christie Jr. as well as other pseudonyms

Because I believed so strongly that Bob Alvarado was the best guy to lead the Regional Council, it was here that I may have accomplished the best organizing of my life. I knew from past campaigns that humor and cartoons could take the boss out. I began writing a newsletter commenting on the state of affairs within the regional council. I tied Casey to General President McCarron, tried to show them as one and the same. I questioned and educated delegates about the Consolidation, pension investments, the failures of recent organizing campaigns, the annuity program, the contract and the "wildcat strike", the efforts to get the vote on the contract back, and the victory of winning back that right to vote.

I tied the efforts by Robert Christie Jr. into the fight in Northern California. And almost all of it was funny. Casey bore the brunt of my sick sense of humor, and as truthful as I knew the situations to be, I took advantage of some individuals for which I still feel a twinge of regret. The cartoons became so widely read that I later heard Casey and his team would spend hours locked in his office worrying about them.

The appointment of John Casey by General President McCarron was at the least controversial, and is covered in greater detail in a previous chapter.

Chapter Twenty-Seven: *The Rank and File Movement*

It is probably here that I should describe the Rank and File Movement in Northern California. In the 1920's a very strong and militant "Rank and File Committee" existed and fought against the American Plan and the Employers Associations. Pile Drivers Local 34 had been known as the "Red Local" during my time. To this day there are communist in Local 34 who paid the price for their beliefs during the 1950's and house on un-American activities trials. Jack Wagner one of local 34's greatest leaders was a communist, they weren't reactionary partisan men, they were fighting for a better world for all working people. Something the communist of today have completely lost sight of, the fight didn't change they have.

The Carpenters Union in Northern California has a long and proud history of Rank and File participation. During the 1920's a Rank and File Federation was formed and working carpenters joined together to right the perceived injustices of the mainstream union.

During the American Plan Strike in 1921 Carpenters from Local 483, spearheaded a Rank and File newspaper and the members of Local 22 voted 5 to 1 against the Building Trades concessionaire agreement and went on strike. Carpenters formed the "Rank and File Federation of Workers of the Bay District". Rank and file carpenter Norman Tallentyre, at a meeting of more the 8,000 workers. Stated "If we cannot go out together, stay out together, and go back together, our last weapon has been broken in our hands, the master class will have nothing more to fear."

The Rank and File Federation had bylaws and formal meetings, even conventions, held in conjunction with other trades who had formed their own federations.

Many carpenters had taken active roles in the "Wildcat Strike of 1974" or were active in the anti-war movement and social unrest that had taken place in Northern California during the 1960's, protesting the implementation of wage controls by then president Richard Nixon.

In fact, this Wildcat Strike had a tremendous effect on many carpenters, including myself, drawing them into their unions and activating them after participating in the strike, their unions, the anti-war movement, and fighting for social justice.

Through-out the 1980's a band of militant Rank and File Members formed the Rank and File Committee to Protest the Two Gates System. They independently selected nonunion contractors picketed and harassed them.

Individual members published a newsletter called the Hardhatt Times, they had a trailer they would pull onto nonunion sites with a two-headed alligator the shot steam out of it's nose. And they could on occasion turn out a few hundred people to protest.

But by and large the majority of those involved were people who had been disenfranchised in one way or another by their local unions. Many of their successes would disintegrate into local union infighting among themselves if any elected leadership involved themselves in their efforts.

At one point in the 1980's the employers unsuccessfully filed suit against many of the Rank and File Committees members, including Donna Levitt, who would become the first women Senior Business Agent of Carpenters Local 22.

One of the members of the Rank and File Committee was a Carpenter named John Riemann. A small man, but very intelligent and articulate, John is a very capable carpenter and worked when ever he wanted. He had been elected to numerous offices at Local 36 and the Local 713 in Alameda County following the mergers of the 1980's. John became a carpenter after moving to the San Francisco Bay Area during the Wildcat Strike over the wage freeze by then President Nixon.

John's vision of the union is a union driven by the Rank and File member and a union run with out paid officials. For as long as I can remember the Carpenters Master Labor Agreement has been voted on by the working carpenters in Northern California. The 33 original bylaws implemented by the International Union did not require a vote by the membership to ratify agreements. And allowed the executive officer of the regional council to appoint the negotiating committee and approve any agreement negotiated by them without membership ratification. In Northern California voting on the agreement by the membership was as American as apple pie. It was inconceivable that someone would even think about taking that right away.

When the councils were merged in 1997, there were thirty-three mandatory bylaws inserted into the NCCRC bylaws by the international union. One by-law gave the Executive Board of the Regional Council the right to approve agreements without a vote of the membership. That meant that 250 elected delegates would decide for 30,000 carpenters which terms and conditions they would work under.

On June 1, 1999 the Carpenters Master Agreement under which the majority of members work was about to expire. The Northern California Carpenters Regional Council under the appointed Executive Secretary Treasurer (EST) John Casey, decided to have the membership respond to a survey, rather than allow the working carpenters to vote on their union contract, something they had done for over one hundred years. Casey and the appointed executive board decided to extend the agreement by approval of the regional council delegates and without a vote of the membership. The survey was perceived to be a trick, to force the membership into a concessionary agreement. The Pile Drivers, who were under the Trusteeship of the International were not allowed to vote on their contract.

The delegates to the Regional Council consisted of approximately 40% full-time employees of the Regional Council and 60% working rank and file carpenters. When the vote was taken 100% of the full time representatives and a small number of rank and file delegates voted 127 to 107 to accept the agreement.

During the meeting a crowd of angry carpenters waiting outside the meeting crashed through the doors and began shouting "Vote No". But the vote passed by a very narrow margin.

The following Monday, May 10th a meeting of rank and file carpenters was held in the parking lot of carpenters Local 713.A crowd of nearly seventy-five carpenters and many carpenters who were working on the San Francisco Airport Extension. They brought up the subject of striking and a vote was taken to move forward. By Tuesday the NCCRC had gotten wind of the strike and sent a representative Tom Delucca to try and stop the wildcat strike.

On May 14, 1999 hundreds of Rank and File carpenters went on strike. They went from job to job with home made picket signs protesting the agreement which no working carpenter was allowed to vote on. Job after job shut down more carpenters joined the march and went on to next job and shut it down. Rough estimates were that as many as 2,000 carpenters walked off their jobs.

The employers went wild, they thought they had a deal and know their jobs were shut down. Carpenters walked off jobs with Project Labor Agreements with negotiated "no strike clauses" shut down. These agreements are negotiated with developers and owners who do not hire carpenters directly. Because they do not hire carpenters directly they do not sign the typical pre-hire agreement contracts sign. The "Project labor Agreement guarantees 100% union contracts on the job and in exchange the union agrees there will be no work stoppages, or strikes by the union. The penalties are very high, as much as $10,000 per day.

Relationships between the union and the Employers grew more strained. The job at the San Francisco Airport and San Francisco Giants Stadium was shut down, the other building trades union were beginning to support the striking carpenters. Leadership of Building Trades Unions began to attack the leadership of the Carpenters. The fight against the International Union over jurisdictional issues spilled over into the "Wildcat Strike", penalties began to accrue at $10,000 per day.

Other Building Trades Unions joined the effort and refused to cross the illegal picket lines. The members of Pile Drivers Local 34 proudly honored the picket lines. When the construction of the foundation of a building stops everything stop.

There were confrontational incidents which pit the agents against the members, who perceived the agents to be siding with the Council. At one demonstration at the Regional Council, Casey ordered representatives out side to confront the crowd while he remained locked inside. These types of actions where Casey and his leaders forced agents to intervene and side against the rank and file drove a wedge between the Casey Team and fulltime staff which became was un-repairable.

This strike inadvertently proved to be the third front in our attempt to save local 34 from non-existence

The Rank and File held rally's in front of the Regional Council offices in Oakland attended by hundreds of working carpenters demanding to speak with John Casey the NCCRC Secretary Treasurer. The main speaker and organizer of these rallies was John Riemann hated by the establishment and feared by the new leadership who had almost all come from outside the "wackie bay area".

The strike ended within a few days, but the point had been made. Carpenters wanted to vote on their agreement. Local 713 and Local 22 sent letters to the International Union that protesting the ratification on the grounds that notice had not been given in a timely manner in violation of the Regional Council bylaws.

International president John McCarron ruled in favor of the two local unions and a second vote was held and the agreement was again ratified by the delegates.

By July 1999, carpenters were striking illegally against a contract they did not vote on. The Building Trades Unions angry with jurisdictional issues in Washington DC. And the Employers pissed off because their jobs were shut down were suing to enforce the Project Labor Agreements.

The issue of voting on the contract was also being fought in many areas around the United States. Finally, a decision was made by the International Union to reconsider the mandatory regional council's bylaws and allow the members to vote on the agreement.

It was during this time that General President McCarron, who was taking a beating nationwide over the by-law, which allowed Regional Councils to negotiate agreements without a vote of the membership. The wildcat strike broke the back of this mandatory by-law. The General President issued a letter to all councils, which waived this by-law and allowed regions to have ratification votes on their agreements.

To my knowledge this was one of the few times the General President caved into pressure by the membership. And we can all thank the rank and file members of the NCCRC for their courage. Many lost their jobs and returned to the out of work list, never complaining or whimpering. A real ratification vote was held and the contract was approved by a majority of the membership and the men went back to work.

International union charges were brought against Riemann by General Executive Board member Michael Draper and a trial was held. Riemann was found guilty of Disobedience of Authority, the authority being the union contract.

John appealed the verdict, and it was rejected by the International union. He then took his appeal to the International Convention held in August of 2000,but was not allowed to present his case on the convention floor.

At the next union meeting at local 713, Tom Delucca the Local Union President informed John that had been removed from office and expelled from the union. DeLucca tried to fill the Recording Secretary position that night by asking for volunteers, no one would take the job in protest to Riemann's expulsion. Within minutes Tom dropped to the floor and was taken to the hospital having suffered a brain aneurism.

While he did survive he had to retire and was never recovered completely. Tom DeLucca was a full time representative who had also participated in the Rank and File Movements of the 1970's and 80's. Like many of our generation, he believed that change had to come from within the organization, but the consolidation of the Regional Council, the fight over the Wildcat Strike and finally the opinion of the membership had been to much for Tom.

The one causality of the "Wildcat Strike" was a carpenter from Local 713 named John Riemann. John was an outspoken leader in the strike, and internal union charges were brought against him. I have known John for almost twenty years; he has always been a vocal critic of the union, and always an elected delegate or officer in local 713 in Hayward Ca.

The charges were filed, a trial was held and at the General Convention in 1999. John Riemann was expelled from the United Brotherhood of Carpenters. Singling out one member for refusing to work and standing up and leading is one great injustice. John Riemann became a martyr to every dissident carpenter in the North America. These two incidents, taking away the members right to vote on their agreement and the expulsion of John Riemann drove a un-repairable wedge between the Official Union and the militant rank and file. I continued to make personal appeals to have the expulsion of John Riemann lifted but to no avail.

By law Riemann could not be refused the ability to work under the union agreement, he just could not hold office or participate in union matters. He continued to work until his retirement in 2007. He may very well be the reason carpenters in the United States continue to vote on their agreements.

This strike was the beginning of the end for John Casey, I began to receive calls from both rank and file activists and disgruntled organizers who had worked for me when I was Director of Organizing for the Bay Counties Regional Council. For years I had been collecting the names and addresses of those carpenters who had taken the Stewards Training Program, and organizers who had worked for me.

Chapter Twenty-Eight: *The Fight for the Northern California Carpenters Regional Council*

Possibilities opened up and opportunities to vote Casey out appeared. But first a credible candidate hand to be found. The only person I believed who could beat John Casey was Bob Alvarado. Bob was the highest vote getter in his local, he had served in almost every position in the organization and he was the person Jim Green had selected to replace him if he had not first been forced out by the General President.

Unhappy agents and organizers began to call me. I began to urge every full time representative who was unhappy about the current state of affairs to call Bob Alvarado and give him their support and to begin to cultivate and identify other staff members to contact Bob and change the union. This first committee of representatives was made up of organizers, agents and apprenticeship staff and others in the organization, including Cliff Dresher of Local 713, Rick Radcliff of Local 751, Dan Digardi of Local 751, Dave Parmerlee, Lana Cantrel of Local 180, Jim MacNamee David Rhodes and Thom Donnelly of Local 34. The wildcat strike was not the only reason many of the organizers and representatives wanted change. Many who had worked for me had little respect for the two organizing directors they were assigned to, largely because my guys had more experience than the people who were their immediate supervisors, and the supervisors' organizing campaigns were failures. In addition, there were obvious differences in treatment by the two directors towards the Bay Counties Organizers as opposed to those from other regional councils.

Each Director had their own department; one department was lead by John Hadzess, a very likeable guy with limited focus. Organizers were spending all their time going on nonunion jobsites, collecting names and addresses of non-union carpenters, handing out bottles of water and carpenter's pencils, but were not allowing any attempts to organize the employers where the men were working. The other director was Richard Wright, a capable organizer with a heavy hand. In December of 2000, organizers under Wright's direction were ordered to picket a non-union job at a mall were a Santa Claus was going to be. A big argument and confrontation occurred and the press portrayed the Union as attacking Christmas, Santa Claus, apple pie and the girl next door. For months following this disaster, organizers in Sacramento were asked if the regional council had plans to picket the Ester Bunny.

By election year the NCCRC needed something to crow about. They signed an agreement with the Residential Builders Association (RBA). Joe O'Donohugh, an Irishman and one of the most cunning and crafty businessmen in San Francisco, ran the RBA. The small residential builders in San Francisco were pressing the city of San Francisco to allow them to build residential units called Live Work Lofts in the areas zoned for industrial use. In exchange for political support the RBA signed with the union.

The city of San Francisco approved the re-zoning and the union got its agreement. It was touted to the members and the general public as though every residential project in the City of San Francisco was now union. But there was slight problem; those who negotiated the agreement waived the subcontractor's clause and about 300 projects that had not even gone out to bid. Upon reading the RBA agreement the members of Local 22 came unglued, and Joe O'Donohugh was delighted to brag publicly how he had screwed the Carpenters Union. The deceit and cover up of the flaws in the RBA agreement was one more nail in the coffin and another reason for change.

At this point Bob had no idea I was moving people to contact him. He is a shrewd politician and would not commit to running unless he thought he could win. I was helping him determine if the support was there. I started list-building the Regional Council Delegates and evaluating support just like I would in an organizing campaign, and wrote a newsletter to those on my list, kind of a gossip column for the Northern California Carpenters. I would receive information, verify it and publish it.

I also began meeting with rank and file carpenters who were delegates, at one point the group became so large that we started holding regular meetings at Machinists Local 1456. Those meetings were attended by many folks, including

Pat Mulligan and Pete Peterson of Local 22, John Jung and Johnathan Noel of Local 713, Joe Holloway of Local1618, Thom Donnelly and David Rhodes of Local 34, Dave Palmerlee of Local 1240, and Joel Vasquez, Matt Kelly, and Noah Teegarden of Local 46, and many others.

Although these were serious men committed to changing the union, convincing them to support Bob Alavardo was extremely difficult. They trusted no full time official and wanted to clean house. They also could count votes and knew that the chance of voting out the current administration was almost impossible. My goal was to win the support from half the full time union representatives and a majority of the rank and file delegates.

In January of 2001 Mary Lieser, journeyman carpenter and apprenticeship coordinator at the Apprenticeship School, announced her intention to run for the position of Vice-President on the Executive Board of the NCCRC.

Richard Wright, one of the Organizing Directors held the vice-presidency. The Casey Team went wild because they thought the campaign had begun and they would also have competition, they began their campaign prematurely, demanding support and money from representatives and organizers alike.

Bob Alvarado told Casey he would not be part of their slate and in fact could not vote for many of the Executive Board members. The following day he was demoted from the position of Central District Manager and moved from his office into a small room with no windows. He was assigned to work on developing outreach programs in the various communities where he could negotiate pre-apprenticeship training. He was replaced by Darrell Thacker as Central District Manager, who was already on the Casey Team and a member of the executive board.

At the meeting where Alvarado refused to sign on in support of the Slate, Organizing Director Richard Wright got into a serious confrontation with Casey. In the middle of the meeting Wright quit his job and stormed out of the meeting. Casey got up and chased him outside and begged him not to quit. Casey's support for Wright, which was becoming a campaign issue, s and his lack of support for Alvarado, who was very popular with the staff, is just another example of the series of bad decisions being made by John Casey.

Now the Central District business agents were angry about Alvarado's demotion and they began to join the fight to remove the incumbent officers. Casey's demotion of Alvarado couldn't have helped our efforts more. Bob became the only possible candidate to unseat Casey. His isolation and new assignment put him in touch with every local union and full time delegates.

I was crunching numbers in January, we believed we needed 60% of the delegates to pledge to vote for Bobby before he would commit to running against Casey. Our past experience told us that beating the incumbent officers would be next to impossible. I knew that once rank and file delegates began receiving threats against their ability to work should they vote against Casey, our chances of winning would be slimmer.

Our numbers continued to grow, and we evaluated support based on a number system - a person was considered a "1" if they had told Bob face to face that he had their support and wore a lapel pin, which was the letter "A" with wings on it. A voter was considered a "2" if they indicated that they might vote for Alvarado, and a "3" if they were a no vote.

By April, 2001, three months before the election, our numbers were excellent; we had commitments from 70% of the delegates. At the April, 2001, meeting Bob Alvarado was confident enough and accepted the nomination for Executive Secretary Treasurer. I knew we could not let up and we continued to campaign. It wasn't until July, 2001, that the Casey Administration realized they were in trouble, but it was too late; Alvarado had personally contacted 70% of the delegates, and I was working with the dissident rank and file carpenters supporting their slate of candidates.

The rank and file carpenters who had supported the "Wildcat Strike" were adamant in their effort to run a candidate against every executive board member. It was like herding cats to convince our rank and file to give their support to full time representatives Alvarado, Mary Lieser, and Dave Palmerlee in order not to split the vote. This took a lot of hard work on the part of these three full time officers, too, because the rank and file carpenters were not absolutely convinced that these full time officials would not turn their backs on them after the election. At the Apprenticeship School there was tremendous support. John Bullock, a long time friend of John Casey, was pressing those instructors who were delegates to support the Casey team. Lana Cantrell, the Administrative Officer, and Jim McNamee, the Trainee Officer from Local 34 were working to win support for Alvarado. They won over the votes of the instructors, and while they were a small group, every vote counted.

I went to work on Local 751, where John Casey was a member and where he should have been strong. Dan Digardi and Rick Radcliff , both organizers from my old department, worked on behalf of Alvarado and we took 75 percent of John Casey's home local.

By the final weeks, the Casey Team was telling rank and file supporters of Alvarado that if they showed up to vote, they would not work. One rank and

File delegate from Redding who said he was voting for Alvarado reported being told this, and it happened over and over. Elections were held to fill delegate positions just under the wire and carefully selected by the Casey Team to have Casey voters at the meeting on Election Day. They also made sure that those alternate delegates who supported the Casey team were eligible to vote on Election Day. Meanwhile the Casey Team was calling agents and organizers in and asking them for $100 to support the Casey Team and calling them in on work time and trying to force commitments from the full time staff. Then Casey made a critical error at the Local 751 picnic in July. Casey asked an organizer, Hans Carlberg, a delegate from Local 751, to put Casey's campaign button on. Carlberg refused and they got into a heated argument in front of a large number of members and staff. Within hours every organizer for the NCCRC had heard about the argument and was angered that Casey was imposing his authority so heavy-handedly; this solidified support from the organizing department even further.

Early in 2001 when Bob decided to run for office, we spoke about how we would run the campaign. I know from personal experience that elections are won by those who make the most contact and have the active and open support of those involved. If I go to bed on the eve of the election believing I have done everything I could, then my chances of success are good. If I go to bed knowing I should have made one more house call, or there was something left undone, then my chances for success are worse.

I told Bob that if he had a chance to win it was slim, that he would have to make personal contact with every eligible delegate, he would have to go to every union meeting, and he would have to be on the phone talking to people until midnight the night before the election. On That evening I called Bob at 11:00 p.m., he had just made his last call and we felt good about our chances.

I didn't have to tell him how to campaign for office, but I stayed on his ass. Tracking the commitments from delegates, it looked like we had a twenty to thirty vote lead on Election Day. Meanwhile I was still winning elections and negotiating first agreements on my real job with the Machinists.

By Election Day our twenty-vote lead dwindled as alternate delegates began to vote, and the threats of job loss worked on enough delegates, who did not show.

I had saved one last bullet for Election Day. During his years in office John Casey had allowed "Letters of Understanding" to be written and attached to agreements signed by the Organizers and Agents. These LOUs amended the Master Agreement in ways largely beneficial to the employers and somewhat detrimental to the men who had to work under them. In these LOUs, for example, the subcontractor clause could be waived, which allowed employers to have nonunion men on their jobs. Another would state employers were allowed to use lower residential pay rates on commercial jobs. Employers were allowed to sign the agreement and bind only one of their companies, which allowed them to operate nonunion. They allowed employers to sign for their commercial work and operate nonunion on their residential work. But the straw that broke the camel's back was the language, Casey had agreed to, in many of these agreements which stated, "Neither the existence of this Agreement, nor any of its terms, shall be disclosed to anyone without the consent of all parties of this Agreement…"

The delegates of Pile Drivers Local 34 were more than happy to hand out the leaflet containing this information following all these months of mistreatment, and they made sure every delegate was handed one on Election Day.

As the delegates voted, David Rhodes of the Pile Drivers challenged John Casey's vote. John Casey was a delegate because he held the position of EST. He was never elected by his local union. Federal law requires that voters for union positions be elected by the membership. Rather than challenge the election committee's decision, John Casey choose not to vote so as not to give Alvarado a reason to challenge the election.

Two of the Millwright delegates, having flown in from Los Angeles just to vote for Alvarado, arrived just in time. Four of the Rank and File carpenters from Local 713 voted for the slate but not for either candidate for EST. The room was spilt fifty-fifty. Bob Alvarado won the position of Executive Secretary of the NCCRC by two votes.

At the end of the day every delegate was sure that his or her vote had been the deciding one in the election. We had fought tooth and nail right down to the wire. In the end, Bob Alvarado had more stamina and will to fight than John Casey. He won along with Mary Lieser, the first woman to be elected to a Regional Council Executive Board nationwide, and Dave Parlmerlee. While no rank-and-file delegates were elected, Alvarado made sure that rank and file representation would be on the executive board. My experience working for the Machinists union had taught me one thing: working members must sit at the negotiating table. The Machinists Union does not negotiate contracts without a committee made up of the workers participating in negotiations.

The day following Bob Alvarado's election we met in my kitchen and I brought this concept of rank and file participation up to Bob and he made it part of his new platform. I am grateful that he listened but even more, it was his ability to listen and implement the idea that proved to the dissident members that he was open and willing to make drastic changes in the organization and recognize the importance of rank and file participation in governing the union. This concept of worker participation is seemingly foreign to the building trades unions except Pile Drivers Local 34. Bob created seats on the executive board held specifically for rank and file carpenters. When the first opening occurred Bob appointed Bill Evans, President of Local 405, to the executive board, after asking Joel Holloway from local 1861 if he would take the position, but Joel declined. In following months, other rank and file carpenters would hold elected seats to the Regional Council Executive Board. Joe Holloway and other rank and file carpenters would be elected to the negotiating committee on the Conference Board, which negotiates the Master Agreements. During the 2003 negotiations, rank and file carpenters sat on both sides of Bob Alvarado.

I was offered the position of Director of Organizing, and gave notice to the Machinists Union. I will be eternally grateful to them for all the support and education I received, in particular Don Crosatto, Joe Bobo and Mike Day.

Bob Alvarado – Courtesy Northern California Carpenters Regional Council

Chapter Twenty-Nine: *Bob Alvarado*

The delegates were going wild at the 1991 General Convention, the general convention was shut down, no one could speak. General President Lucassen had refused to allow Whitey Rodgers to speak; he could not stop the delegates from pounding on the tables in protest.

Some one had handed out straw hats to every delegate, the hats were being used like drums, some delegates were wearing the hats on their arms, after they had punched there fist through the tops. Others were screaming at the tops of their lungs, no one could speak.

I got on the walkie talkie and asked the "Whips" to take the noise level down; they could not silence the crowd. The noise still did not stop. Each Microphone on the floor of the convention had a number above it. The Northern California delegates were screaming at the podium to recognize microphone nine.

I called to John Walsh, one of our delegates, to run up to the podium and tell Lucassen to recognize microphone number nine.Walsh charged to the front of the room and franticly waved to Lucassen, who stood dumb struck like a deer in the headlights. Walsh screamed, "You want this to stop, recognize microphone number nine".

Lucassen looking up franticly at microphone number nine saw the delegate standing there with a bright pink baseball hat on ready to speak. The bright pink hat signified that the speaker was a Lucassen supporter. Unknown to him Bob Alvarado had been instructed by Jim Green to "procure one of those pink fucking hats and park your ass on microphone number nine."

Bob and another delegate from the Bay Counties, Donna Levitt, had stopped outside the convention center and pretended to switch sides, trading both of their "Whitey Rodgers" buttons for pink hats the Lucassen supporters had been wearing. Donna Levitt had gotten her hat from Gary Martin a delegate from Monterey Ca. who had switched sides.

Lucassen yelling into the microphone "The chair recognizes the speaker on microphone number nine", we got the delegates to quite down. Microphone number nine was switched on. "Mr. President my name is Bob Alvarado from Carpenters Local 152 in Martinez California; "I relinquish my time to the previous speaker John Whitey Rodgers".

The convention center went wild; the Whitey Rodgers supporters were dancing in the isles, rolling on the floor in hysterics. The "Siggies" looked as if someone had punched them in the stomach. Sig Lucassen stared around the podium, like a man looking for a fire extinguisher. He looked down at the microphone and said, "The chair recognizes Whitey Rodgers, the room erupted.

Bob was born in Long beach California in 1953. With his mother and brother they moved to Port Costa California, where she married Rudy Alvarado. He graduated from John Swett High School in 1970 and was apprenticed into Carpenters Local 642 in 1972. Alvarado was a residential carpenter, a "framer". For the next fourteen years he worked in every aspect of homebuilding, from apprentice to journeyman to foreman, and when the company decided to do more commercial work Bob was the first choice for the job as superintendent. He was elected President of Local 642 in 1981 and served in that position until local 642 was merged with Carpenters Local 2046 and became carpenters Local 152, where he was elected President and served until 1998.

In January of 1985 he applied for the position of Organizer for the Bay Counties District Council of Carpenters. During those days there was a dress code which required all organizers to be clean shaven. Bob had noticed that other staff members at the council wore beards and did not think to ask if his beard would be a problem. Roy Fouche, the Director of Organizing at the time was very "old school" in that he had rules and expectations regarding dress and appearance. After being hired, Bob was informed that he would have to shave his beard. He refused to shave and declined the job. To some observers at the time, it seemed that shaving was an irrelevant event...what was most important was whether he wanted the job or not; was his beard more important than working for the union?

However, in retrospect, this act showed his true conviction to his principles. He had not been told there was a "no beard policy" and to shave his beard off would require him to give in on an issue that had not been agreed upon from the start of the hiring process. This conviction to honoring an agreement and standing up to the consequences would follow him throughout his career.

Like most people I have met in the labor movement, those who move up in the organization have a mentor, I call them a "rabbi", someone who sees talent or special ability in them and who works with the person to develop it. Some people have more than one. Bob's "rabbi" was Scott Bournonville, the Financial Secretary of Local 642. Bournonville was small in stature but as brave and smart a man that I have ever met. His greatest trait was his ability to fight for what he believed to be true or the right course of action for his local union. If his position was not approved of by a vote of the membership, he could move on, and fight for the next issue without anger or animosity and with the same fervor as before. He was truly democratic and believed he represented the majority of members at his local union. Until his retirement in 1996, Bournonville gave good counsel and training to Alvarado.

Jim Green had introduced and had the membership approval of the "dues check-off", a new way of financing the district council and local union's operation. A few local unions who could afford to hire an organizer to work directly for the local union did so. Bournonville convinced Bob to apply for the position, and after the membership voted to hire him Bournonville mentored this sometimes hardheaded carpenter in the art of local union politics. He also allowed Bob to keep his beard and cowboy boots!

It was during this period of 1986 through 1988 that the local union mergers were taking place. As president of local 642 Alvarado oversaw the merger along with the President of local 2046. Even though the merger had been agreed upon by the locals involved, the transition between unions can be very difficult and complex, with many issues to be resolved, such as how to handle the Local's union funds, secretarial staff, buildings and properties owned by the union. The question of who was going to hold union office until elections could be held was a purely political matter that confounded many organizations, throwing them into a quagmire of internal fighting and indecision. Local 642 was a smaller union, financially, geographically and membership wise. While Local 2046 was a larger organization it was more political, with factions that kept the union disrupted and constantly agitated by internal fighting and political maneuverings by the business agents and secretaries. Alvarado, with Bournonville's help, guided the transition to a successful conclusion, concluding with him being voted in as president and Anthony Viola as Financial Secretary of Carpenters Local 152. A very important lesson here is that a smaller organization, with more dedicated and united members can take over a larger organization that is divided by personal ambition and factional infighting.

Local 152 is located in Contra Costa County and represents carpenters who build practically every type of commercial building, high rise, highway, refinery, steel mill, chemical plant and residential home. A business agent in that area must understand many different types of construction and administer that master agreement fairly among the many diverse work types.

With the consolidation of smaller, specialized local unions into larger more diverse locals, friction sometimes developed among the different types of carpenters. Housing industry issues were very different from refinery issues, and dry walling is different than concrete forming. These issues would come to the floor of the local union and where handled by Alvarado.

In turn he learned to administer and negotiate with some of the county's largest and most powerful corporations. As time passed, he became the person the members and employers trusted because he honored his word and commitments. In 1994 he was appointed by Jim Green as the Senior Business Agent of Carpenters 152. In addition, Bob learned to deal with politicians, and in many ways he changed the political model within unions, where politicians would dictate the agenda regarding political issues within a county along with the Building Trades Council.

The Building Trades Council in Contra Costa County, like most Building Trades Councils, was afraid of the carpenters union. The Carpenters, Laborers and Operating Engineers represent the large majority of construction union members. Building Trades Councils usually have a voting structure that does not allow the larger unions to out-vote the smaller unions. Generally the leaders of the Building Trades Councils are from one of those smaller, less powerful organizations. These smaller unions lack the resources to organize workers, so it benefits them to rely on local politicians to do their organizing for them. They become beholding to politicians for help and drain resources that could be used for organizing into the never-ending request by politicians for campaign money.

Local 152, with Bob as president, began to channel money into local political campaigns in Water District races, School Boards or agencies where new people starting out in politics needed help and could be cultivated as loyal and true friends of the Carpenters Union. Many times Local 152 bypassed the wishes of the Building Trades Councils and their endorsements, choosing instead to support their own candidate.

In 1995 Jim Green, the Bay Counties District Council of Carpenters Executive Secretary, decided to retire and asked Bob to leave local 152 and come to work at the district council as his Administrative Assistant. I had noticed that over the last two years Jim Green had begun to show signs of stress and had on more than one occasion stated that he wanted to retire. By choosing Alvarado in 1995, he could set a date to transition the district council from his control to Alvarado.

Jim Green began to turn the day to day running of the union over to Alvarado. He put him in situations to learn the ropes. Green told employers and union that if Bob Alvarado spoke he was speaking for Jim Green. Jim guided him and trusted him and watched his back as he had done for me. During the next two years Bob learned how to run a district council.

When the international told us of the upcoming consolidation, we had all been assigned responsibilities mine was to put together an organizing plan for a new Regional Council. Others had their assignments; it was Alvarado's job to put it all together. Green had anticipated the consolidation and we all assumed he would head the newly created Regional Council. It was also assumed that following the consolidation Alvarado would replace Green as the Regional Council Secretary.

That was not to be, John Casey was appointed Executive Secretary of the Northern California Regional Council of Carpenters by Doug McCarron. I quit and Bob went to work as the Central District Manager.

As time passed and the membership grew more dissatisfied with Master Agreement negotiations, the rank and file wildcat strike, and the ongoing problems the Pile Drivers were creating. Casey allowed the perception that the International was controlling every aspect of union business. For these reasons Alvarado decide to run for the position of Regional Council Secretary. And because the Carpenters Union is the lead craft on any job and throughout the history of the Building Trades Unions in Northern California. Whatever issues the carpenters union tackles, other trades follow, whether it is wages, benefits, negotiations, and organizing; the carpenters usually set the pace.

Under John Casey the relationship between the Carpenters Union and the employers deteriorated. Politicians no longer looked to the Carpenters Union to set the standards that had existed between them and Jim Green and the Bay Counties District Council. The master agreement negotiations failed to improve standards and instead conceded hard won benefits during a climate of booming construction projects. And finally, the rank and file were losing faith in the organization and were in open rebellion with the Regional Council, a fact that did not go unnoticed by the employers.

Alvarado wanted to return the union to its former self and improve on what had been started by the delegates of the Bay Counties District Council in 1978.

In hind sight, the appointment of John Casey may have been better for Bob Alvarado. If he had inherited the position from Jim Green he would have also inherited political debt that Jim Green carried.

You see Jim Green was the consummate deal maker. To succeed in the initial mergers he had made many necessary deals and arrangements with local union executive boards, business agents and members and the International. John Casey had similarly had to make deals and uncomfortable bargains with his contemporaries to secure the appointment as the Regional Council Secretary.

The appointment by General President McCarron cast Casey as someone who was a puppet of the general president, who had not voice or opinion of his own. Casey dealt with the consolidation harshly and heavy handedly. The rank and file never appeared to accept John Casey, for these reasons, in my opinion.

I believe Bob Alvarado would still have succeeded had things gone the way we had planned, but by waiting and winning the position in a hard fought election, I believe he avoided being saddled with the political debts of both Jim Green and John Casey.

As Executive Secretary of the Northern California Carpenters Regional Council he has diversified the trust funds, limiting risk while other union pension funds have dried up. He corrected the deficiencies in our Health and Welfare Plan while the rest of the nation struggles to provide adequate health and welfare for its members. He has regained the respect of employers and politicians, making him quite possibly the most powerful labor leader in California. Upon taking office he returned the right of the membership to vote on the Master Agreement. He requested from the delegates that rank and file carpenters sit on the Executive Board of the Regional Council and the Master Agreement negotiations, and he regained the trust of the membership.

However, in my mind his greatest strength is his charisma and his willingness to always stand up in front of the membership and explain difficult matters in person. Early in his first term our health and welfare was in trouble. Actuaries the union paid big money to were recommending health and welfare cuts. Rather than accept their recommendations he had the health plan broken down item by item and savings were found without making dramatic cuts in benefits. And the actuary was fired.

He then went local by local and stood before the membership and explained the benefit plan and took questions from everyone who wanted to speak. He has done this time and time again.

His success in office was by design, the cream rising to the top, a system created by working carpenters in 1978 who believed in the union, and which evolved, albeit painfully, into one of the most powerful labor unions in America.

Chapter Thirty: *Conclusion*

My journey began when I was very young; my understanding of unionism was taught to me at the dinner table, and it was personal. Both parents were union members as well as my grandparents. My high school education was vocational; I was identified by a functioning educational system that prepared me for the vocation I chose.

I knew the first day on the job that I wanted to be a "pilebutt", and I chose to participate in union functions, because the men I met were truly men to me and completely inspirational. They were flawed, as we all are; I accepted that from the start.

And as I grew older and gained experience, I accepted that fact that even great leaders are flawed human beings. I asked myself many times when I was a younger man, how did they know that the decisions they were making were the right ones? I now know and believe wholeheartedly that the leaders I knew were making the right decisions because they understood the subjects they made decisions about. They went through soul-searching agony, knowing they held the future of men, women and their families in their hands. And that they believed in their vision of the union membership, the organizations they represented and the larger vision of the International Union.

That doesn't mean that all leaders are prepared to be leaders. Many are ill-equipped to lead or lack the ability In order for the labor movement to revive itself, to save the American worker and possibly the entire middle class of the United States of America, it needs radical change, from the very top to the bottom because workers want unions, and the middle class in this country cannot exist without them. The employers, who are only trying to make more money, and who are taking advantage of the imbalance of our trade agreements, are solely dependent on union busting consultants and the corporate-controlled media, who daily force-feed antiunion drivel down workers throats. And union leadership in a majority of unions seems content to continue the status quo as long as their coffers hold enough cash to keep them in office for another term.

Meanwhile, rank and file members sit at home complaining about their conditions, accepting the corporate drivel about unions and remain inactive and complacent. Any person who seeks public or union office is branded as a sellout. My question is…a sellout to what? The system operates the way that it does, and folks who refuse to take part often just lose out! This attitude makes it seem sometimes that Americans are not allowed to have working class heroes. Meanwhile, corporations continue to lie, steal and cheat Americans; they use their power to undermine support of issues that would benefit all Americans for the sake of their stockholders and profits.

The Northern California Carpenters history shows that when unions, under the guidance of more than one dynamic leader have existed at anyone time, the conditions of the working class have risen. It was during the years of leadership of carpenter P.H. McCarthy that the San Francisco Building Trades grew in power. Or for the Carpenters Union it was during the terms of office of Chester Bartalini, Jim Green and Robert Alvarado. I am sure I will antagonize many of my friends when I say that Douglas McCarron's leadership, while heavy handed at times, has over all served the UBC. As union membership in other trades and occupations takes drastic declines, the United Brotherhood suffered less and remained a steady force. The fact that he is alone in his position at the top of the labor movement proves my point as union membership slides closer and closer to zero.

There are many reasons for the lack of growth in union membership in general, and in our area, building trades' membership in particular. But the major difference between the Carpenters Union and the Building Trades Unions has been the commitment on the part of the organization to change.

This wasn't done easily, as I described the changes made by the Northern California Carpenters Union; these radical changes have been difficult. They took place because one man, Jim Green, had a vision and the moral will to see the changes through. The business agents in 1978 knew that the Carpenters Union would change or die, and the majority supported Green's vision. And the rank and file delegates carefully supported the need to restructure and change the direction of a stodgy old patriarchal organization into one that removed barriers, made work accessible and fairly distributed to everyone and finally change from a tired organization into a union committed to organizing with the support and resources that that effort required. Northern California has a great history of progressive unionism in general, and for us carpenters, at many times during our history everyday working men and women, rank and file members organized to challenge leadership in particular. The poorer the leadership the larger the rank and file movements became. As change began, members who had been active rank and file unionists, deemed radical, went to work inside the organization, and the face of the union leadership changed; African Americans, Hispanics, Asians and others, men and women took leadership roles in the organization.

Furthermore, I believe union educational programs should and need to exist, open to all members--designed to teach rank and file working men and women how the union is structured, how the trust funds operate, what the contract states and letting them get to know the leadership, putting a face to a name they usually only see on letters written by lawyers. When fostered and supported, apprenticeship and journeyman training alike takes off in new directions, giving members opportunity to work and learn every day. I honestly believe that an uneducated union member does more harm to an organization than any radical unionist who understands the inner workings of his or her union. More importantly, union leadership teams need to ask union members to be accountable, and to do something. I was seldom disappointed and many times astonished and pleased when a worker or group of workers would take over an organizing campaign as their own.

This book is partially a history of the UBC during a short period of time from 1978 until 2008. It is also a story of how different men and women changed an old, top-heavy, entrenched organization. You can come to your own conclusions as a member whether or not it is working and who for. I certainly didn't write this book to please one faction or another. I wrote it because the American labor movement needs to change. I wrote it for the union members who believe in unions, and want their unions to exist. I wrote this for union members who believe in unions and want to modernize the things that need change yet hold on to the traditions that are dear to the organization and to encourage rank and file members to dive in to volunteerism and service and not be afraid.

The changes were difficult and painful, and they took years to complete. In the case of Northern California, it was because the majority ruled, plain and simple. Those who did not want change still believed in the democratic principle, that after the majority speaks, we move on to the next issue.

Change must be for the betterment of the organization and not you personally. Change comes by reaching back, holding out your hand and lifting those behind you up with you. Change comes when those you lifted up replace you someday, with better skills and different dreams. Change comes when you reach your own level within an organization and are smart enough to realize you have gone as far as you can and be satisfied with that.

And, change comes from the realization that you are doing this for free, and that your efforts may go unnoticed. That you are working for the good of the organization as a whole, which includes your political adversaries.

And finally, the changes you envision may take 30 years to come to pass and for you to visualize all the good that comes of them, whether or not you are at the forefront of the organization at the time.

Occupy.

Archie Green – Courtesy Bill Myers

Acknowledgments

Adam Machedo
Aaron Seward
Al Roosma
Alan Linder
Alex Kothny
Andy Slivka
Angelo Balistreri
Annika Felton
Anthony Banks
Anthony Ramos
Antoine Rigmaiden
Archie Green
Augustine Beltran
Augustine Diaz
Babe Garcia
Bajun Mavalwalla
Bert Tycer
Bill Evans
Bill Feyling
Bill Graziose
Bill Hopkins
Bill Martin
Bill Meyers
Bill Reeves
Bill Scanlon
Bill Sweaney
Bill Tinsley
Bill Wong
Bob Alvarado
Bob Blake
Bob Borck
Bob Klein
Bob Knight
Bob Pieto
Bob Smith
Bruce Lanham
Bruce Mac Dougal
Bud Bryant
Calvin Watterson
Carl Anderson
Carl Erickson
Charles Carpenter
Charles Hudson
Charles Lambert
Charles Sutter
Cheryl Jackson
Chester Chan
Chris Eldham
Chuck Cannon
Chuck Taylor
Chuck Wagner
Clarence Klein
Cliff Drescher
Cliff Edwards
Cory Strack
Craig Oleson
Craig Strehlo
Curtis Kelly
Curtis Morgan
Dan Digardi
Dan Douglas
Dan Hannan
Dan Huff
Dan Moore
Dan Murray
Daniel Becco
Daniel Murry
Darrel Thacker
Daryl Dickenson
David Imus
David Parmerlee
David Rhodes
Davy Neal
Deb Hientzman
Dennis Baca
Dennis Christensen
Dennis Dean
Dennis McConnell
Dennis McWhorter
Dennis Zampa
Deryle Adams
Dick Olsen
Don Clauser
Don Crossato
Don Dolly
Don Selvey
Donald Tallerico
Donna Levitt
Doug Downie
Doug Whipple
Duke White
Ed Delgado
Ed Yee
Emerson Franks
Eric Taylor
Foy Sargent
Francisco Nieto
Frank Castiglioni
Frank Crim
Frank Mendibles
Frank Thomas
Fred Devine
Fred Esson
Fred Wright
Gary Bakke
Gary Ross
Gary Slauson
Gary Woodworth
Gene Krautheim
George Barreras
George Gunderson
Glen Skogen
Guillermo Hernandez
Guillermo Molina
Harry Allen
Jack Melton
Jack Wagner
Jack Wallace
Jackson Felton
James Crotty
James Fernandez
James Garland
James Halloran
James Lawson
James Walker
Janet Staples
Jaun Gonzales
Jay Streets
Jerry Connors
Jesse Jaurez
Jesus Rangel
Jim Green
Jim Howard
Jim Johnason
Jim Larsen
Jim McNamee
Jim Walker
Joanie Murphy
Joe Bobo
Joe Hart
Joe Holway
Joe Lumsden
Joe Peters
Joe Richardson
Joel Vasquez
John Jung
John Anzini
John Bullock
John Calloway
John Casey
John Coccimiglio

John Finnegan
John Gustafson
John Hillsman
John Jung
John Liguori
John Lynch
John Mattera
John McGuinn
John Riemann
John Walsh
John Wilkenson
Johnathan Noel
Judy Gatewood
Ken Maderazo
Ken Peacock
Kenneth Carty
Kenneth Lukas
Kenny Viscovitch
Kieth Foster
Lana Cantrel
Larry Wagner
Laura Roland
Lawerence Ahern
Lenard Smith
Leonard Ramos
Leslie Munoz
Lloyd Larsen
Luis Solares
Lynda Bogue
Manuel Flores
Manuel Sierra
Maria Brooks
Mariann Hassler
Mark Sutter
Mary Lieser
Matt Kelly
Michael Carter
Michael Day
Michael Lacy
Michael Leong
Michael Orfelt
Michael Roger
Michael Roscelli
Michaelangelo
Krango
Mike Berg
Mike Carvalho
Mike Cassado
Mike Knab
Mike Krautheim
Mike Papalia
Mike Stevens
Ned Van Valkenberg
Neil Ferguson
Niel Breiling
Noah Teegarden
Pat Carvalho
Pat Karinen,
Patrick Dennis
Patrick Mulligan
Paul Cohen
Paul Cohen
Paul Knudson
Paul Lewis
Paul Makala
Paul Norris
Pedro Mendez
Pete Peterson
Pete Schantz
Phil Daniels
Phil Giest
Phil Hennessey
Phylis Clima
Ralph Corral
Ralph Rubio
Ray Cordeiro
Ray Graham
Ray Silva
Red Poulson
Rex Atwood
Richard Fernnandez
Richard Hiebert
Richard Noar
Richard Radciff
Richard Wright
Rick Bonilla
Rigo Liguardia
Robert Baldini
Robert Fouche
Robert Lanham
Robert Lanham
Robert Mattacola
Robert Rooke
Ron Cheshire
Ron Langston
Ron Petingil
Ron Petty
Ron Spikes
Roy Fouche
Ruben Lopez
Sandy Benson
Scott Bournonville
Sean Hebard
Shawn Leonard
Shawn Smith
Steve Campos
Steve Ferini
Steve Grannis
Steve Rigmaiden
Steve Tarabini
Steve Tilton
Strig Liljedhal
Ted Mesinas
Ted Sanderson
Ted Strong
Terry Callan
Terry Price
Theodore Strong
Thom Donnelly
Thomas Deluca
Thomas McCarthy
Thomas Quinn
Tim Elsken
Tim Felton
Tim Lipscomb
Tom DeLucca
Tom Holman
Tom Mattos
Tom Pearl
Tom Post
Tony Castillo
Vernon Houck
Victor Van Bourg
Warren Stevens
Wilbur Peet
Willam Bergerson
William Krebs
William Lyles
William Miner
William Murphy
William Napier
William Scanlon
William Sokol
William Strange